Energy and Transport Processes

Core Programming Topic at the 2011 AIChE Spring Meeting and 7th Global Congress on Process Safety

Chicago, Illinois, USA
13 – 17 March 2011

ISBN: 978-1-61782-901-7

Printed from e-media with permission by:

Curran Associates, Inc.
57 Morehouse Lane
Red Hook, NY 12571

CURRAN ASSOCIATES INC. proceedings.com

Some format issues inherent in the e-media version may also appear in this print version.

Printed by Curran Associates, Inc. (2011)

AIChE
3 Park Avenue
New York, NY 10016-5991

Phone: (203) 702-7660
Fax: (203) 775-5177

www.aiche.org

Additional copies of this publication are available from:

Curran Associates, Inc.
57 Morehouse Lane
Red Hook, NY 12571 USA
Phone: 845-758-0400
Fax: 845-758-2634
Email: curran@proceedings.com
Web: www.proceedings.com

TABLE OF CONTENTS

(6a) Upgrading Low BTU Fuels to Reduce Emissions and Enhance Performance In IC Engines

Natali Y. Ganfer and Marco J. Castaldi, Earth and Environmental Engineering, Columbia University, New York, NY

Landfills are currently the most common system used for the final disposal of solid waste. They are projected to remain the most common system in the forthcoming years. The release of Landfill Gas (LFG), mainly composed by CH4 and CO2 in an average proportion of 50%-50%, to the atmosphere constitutes an important environmental issue as methane is a GHG 20 times more powerful than CO2 and 23% of the methane emissions in the US come from landfills. On the other hand, LFG can be a free source of low BTU fuels. For these two reasons, LFG is in most cases being collected, and it can be used as IC engine fuel in order to generate electricity. However, because of the fluctuations in the composition of the gas, higher emissions of pollutants such as UHC, NOx, and CO are released. Thus the importance of lowering the emissions of these toxic gases whose expulsion is regulated by the EPA. One solution to this issue is the addition of Synthesis Gas (Syngas) -H2 and CO- to the input LFG.

Preliminary data has shown that by supplementing the input gas stream with as little as 5% Syngas, Nox emissions can be reduced up to 40%, CO up to 67%, and UHC up to 80%. Both the amounts of CO and UHC are reduced because the Syngas increases the activity of the fuel due to its higher flame speed; t he Nox formation is mitigated as a consequence of a more stable combustion at lower temperatures. More detailed data needs to be collected for different Syngas, LFG, and total input gas compositions. All the tests are being run in a Honda Engine GC160 with Maximum Power output of 2kW. Nevertheless, emission levels of 4 different engines with bigger capacity-Capstone CR200, Solar Centaur 40, Caterpillar G3516 LE-have been compared. It was observed that the results of the experiments can be extrapolated to an industrial scale as well.

See more of this Session: Thermal Processes for Waste Utilization
See more of this Group/Topical: Topical 9: Waste & Biomass Valorization

(6b) A Technoeconomic Analysis of Biodegradable Polymers Production From CO_2 and Waste Plastics

Michael Hartmann, Maoqi Feng, Thomas Andrews and Eloy Flores III, Chemical Engineering Department, Division of Chemistry and Chemical Engineering, Southwest Research Institute, San Antonio, TX

This paper evaluates the economic viability of manufacturing biodegradable polymers (mostly polycarbonates) from CO_2 and waste plastics. Waste plastics recycling are beneficial to environment and economy as well. Thermal processes have been used for feedstock recycling from waste plastics, such as polyethylene, polypropylene and polystyrene. Olefins are very useful feedstock in industry; however, it's expensive to produce olefins by using conventional hydrocarbons cracking technology. In our simulation, olefins are generated from pyrolysis of waste plastics with thermal energy supplied from a solar concentrator, thus no net CO_2 is produced in the pyrolysis. Biodegradable polymers are then produced from the polymerization of CO_2 and olefins directly in a catalytic reactor. This method provides an alternative method for CO_2 sequestration and utilization. A detailed economic analysis using VMGSim software is presented.

See more of this Session: Thermal Processes for Waste Utilization
See more of this Group/Topical: Topical 9: Waste & Biomass Valorization

(6c) Modeling of Medium Constituents for Optimized Production of Lipase From Niger Seed Oil Cake (Guizotia abyssinica) Using Artificial Neural Networks and Genetic Algorithms

Sarat Babu Imandi, Department of Biotechnology, GITAM Institute of Technology, GITAM University, Visakhapatnam, A.P., India and Hanumantha Rao Garapati, Department of Biotechnology, Anil Neerukonda Institute of Technology and Sciences, Visakhapatnam District , AP, India

A hybrid system of Artificial Neural Networks (ANN) and Genetic Algorithms (GA) was adopted to optimize the medium constituents to enhance the lipase production by *Yarrowia lipolytica* NCIM 3589 in solid state fermentation (SSF) using Niger seed oil cake as the substrate. Different microbial metabolism regulating fermentation variables (initial moisture content of the substrate, glucose, and urea) were studied to construct a '3–5–1' topology of the ANN for identifying the nonlinear relationship between fermentation variables and the enzyme yield. ANN model was subsequently optimized using GA for obtaining maximum lipase production. The usage of ANN–GA hybrid methodology resulted in a significant improvement (25 %) in the lipase yield when compared to preliminary investigations. MATLAB 7 was used for implementing the ANN–GA methodology and the corresponding program was included in the appendix.

See more of this Session: Thermal Processes for Waste Utilization
See more of this Group/Topical: Topical 9: Waste & Biomass Valorization

INFLUENCE OF ACIDS AND ALKALIS ON CELLULOSE PYROLYSIS PATHWAYS AND KINETICS

Salim M. Shaik [(a)], Paul Nicholas Sharratt [(a)] and Reginald B.H. Tan [(a,b)]

(a) Institute of Chemical and Engineering Sciences, Singapore 627883, Singapore,

(b) Department of Chemical and Biomolecular Engineering, The National University of Singapore, Singapore

1. Introduction

Cellulose pyrolysis has been the subject of many studies due to it being the largest component of most biomass sources. A phenomenological description of cellulose pyrolysis was hypothesised by Mamleev et al. [1]. The main thermal degradation pathway is via intermolecular transglycosylation reactions within the glucose monomers of cellulose [1,2]. Apart from transglycosylation, formation of liquid tar from cellulose can also occur via β-elimination. Under this mechanism, volatile acids (e.g. carboxylic acids) formed from the initial cellulose decomposition are able to attack the remaining cellulose as Bronsted acids thus catalysing heterolytic (ring-opening) reactions.

In this current work, we have studied the effects of acids (H_2SO_4, H_3PO_4 , H_3BO_3) and alkalis ($Ca(OH)_2$, NH_4OH) on the yields of levoglucosan and its derivatives (levoglucosenone, 1,4:3,6-dianhydro-α-d-glucopyranose, anhydroglucofuranose, hydroxymethyl furfural, furfural and levulinic acid). A proposed mechanism adapted from previous workers [3,4,5] and based on our current results is illustrated in Figure 1.

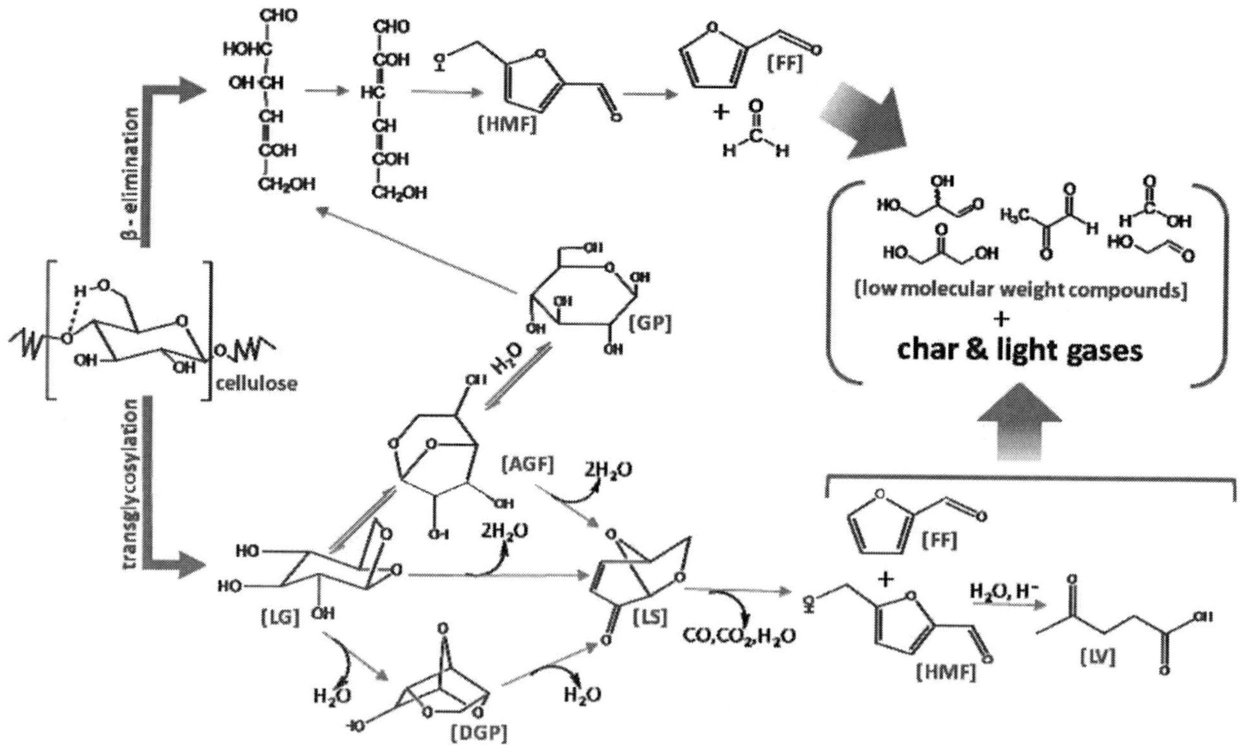

Figure 1: Schematic of possible mechanism for the conversion of cellulose to levoglucosan and its derivatives

In addition to the hypothesised role of volatile acids in pyrolytic chemistry, it has also been known that acid/alkaline content has a significant impact on the yields and product composition during direct liquefaction. Budarin and co-workers had produced bio-oils via microwave pyrolysis of wheat straw [6]. At the same time, they made use of acidic (HCl and H_2SO_4) and alkaline (NH_3) additives to effect changes in the product distribution and composition of the bio-oils derived from wheat straw pellets. The levoglucosan yields of acid and alkaline treated biomass compared with untreated biomass are shown in Table 1 below.

Table 1: Comparison of the effects of acids and alkalis on biomass pyrolysis yields.

	Budarin et al.			Hassan et al.				
additives	H_2SO_4	HCl	NH_3	H_3PO_4	H_2SO_4	NaOH	$Ca(OH)_2$	NH_4OH
levoglucosan	↓ 26%	↑ 27%	↑ 66%	↓ 54%	↓ 61%	↓ 49%	↑ 72%	↑ 41%

Budarin's results were rather mixed with no clear discernable trends between acidic and alkaline additives. A similar trend is also observed based on the results obtained by Hassan et al. [7] who studied bio-oils from pyrolysed pine wood. However, we note that the use of acids (H_2SO_4 and H_3PO_4) decreased levoglucosan yields whilst the alkalis ($Ca(OH)_2$ and NH_4OH) seem to increase levoglucosan yield. Superficially, this seems to provide some evidence for Mamleev's mechanism which suggests that acid catalysis would activate β-elimination whilst the presence of alkalis should suppress β-elimination and promote transglycosylation.

Mamleev's scheme is a novel representation of the reactions involved in cellulose pyrolysis. Although it shares some common features with other models like the Broido-Shafizadeh and Shafizadeh-Chin models, Mamleev seems to be the first to explicitly incorporate the role of acid catalysis in cellulose pyrolysis. Mamleev's scheme, indicated that there is an acid-catalysed (β-elimination) pathway with a higher activation energy (Ea ≈ 250 kJ/mol) compared with the transglycosylation pathway (Ea ≈ 200 kJ/mol) during cellulose pyrolysis. Based on kinetics work done by Banyasz et al. [8] and Capart et al. [9], it was shown that the acid-catalysed, (β-elimination) pathway was favoured as temperature increases (illustrated in Figure 2).

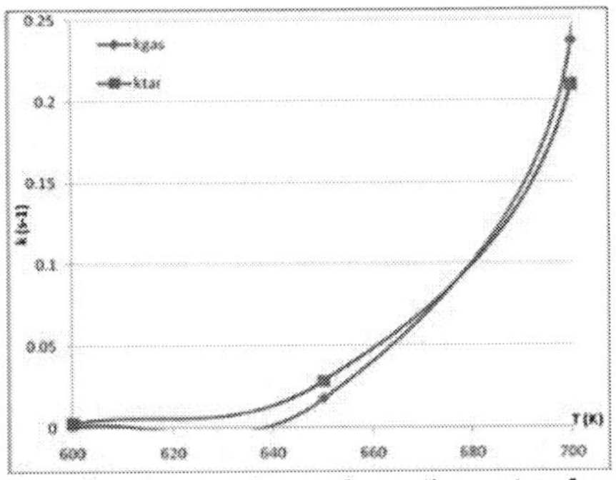

Figure 2a: Comparison of reaction rates for β-elimination (k_{gas}) and transglycosylation (k_{tar}).

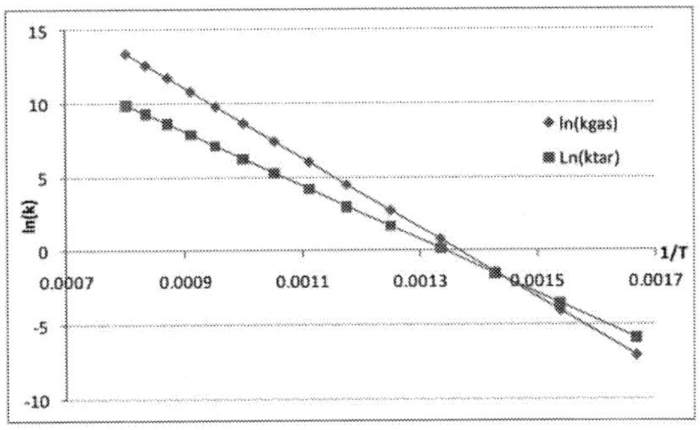

Figure 2b: Comparison of ln(k) against reciprocal temperature.

The graphs in Figure 2 also indicates that the "cross-over" temperature from transglycosylation dominated pyrolysis to ring-opening (β-elimination) dominated pyrolysis occurs at circa 670 K.

The formation of levoglucosan is expected to occur solely via the transglycosylation mechanism. It is therefore expected that the yield of levoglucosan would decrease with increasing temperature and acidity.

2. Experimental Method

The experimental setup used to study levoglucosan yield from cellulose pyrolysis consists of a 1.5 inch diameter stainless steel tubular reactor with a sample holder (capacity ca. 15cm^3) placed inside a horizontal furnace (Carbolite). The furnace is capable of heating the reactor up to 1200 °C and at a rate of 20 to 30 °C/min. Sweep and/or reacting gas is provided via two gas flow controllers which enable variations in gas compositions and reactor residence times. For the current work, nitrogen flow rates circa 650 ml/min were used and this translates to a residence time of 2 to 3 seconds.

Volatile tars and gasses that leave the reaction zone are allowed to cool immediately via three condensers (cooled to ca. 20°C, 0°C and -20°C). The condensed volatiles were analysed using a GC-MS (Agilent: 7890A GC, Column: DB-1701, 60 m x 0.25 mm I.D., 0.25 µm) for levoglucosan and anhydrosacharides content.

Cellulose samples were prepared at different acid and alkali loadings by stirring microcrystalline cellulose (Avicel from Sigma-Aldrich) with the required acid/alkali solution (0.5 to 5wt%) at ca. 25°C for 30 minutes. The "wet" cellulose samples were then filtered and dried overnight in an oven at 60°C.

3. Results and Discussion

The various components of interest namely levoglucosan **(LG)** and its derivatives (levoglucosenone **(LS)**, 1,4:3,6-dianhydro-α-D-glucopyranose **(DGP)**, 1,6-anhydro-β-D-glucofuranose **(AGF)** were identified via their mass spectrum (NIST library) and their retention times The relevant data for the standards are listed in Table 2. Quantification of the various species was done with naphthalene as the internal standard.

Table 2: Identification parameters for the main pyrolysis products

Component	LG			LS			DGP			AGF		
Retention time (min)	22.9			17.5			18.8			24.6		
Main mass fragments (m/z)	60	57	73	98	96	39	69	29	57	73	69	44

From the chromatograms, the yields of levoglucosan and its anhydrosacharide derivatives obtained from pyrolysing cellulose with varying acid/alkali loadings were calculated and compared as shown in Figures 3a to 3e.

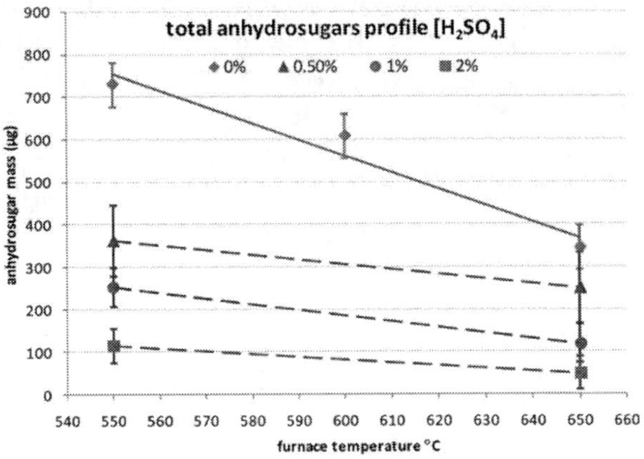

Figure 3a: Anhydrosugar yields from cellulose treated with varying concentrations of H_2SO_4 (0.5wt%, 1wt% and 2wt%)

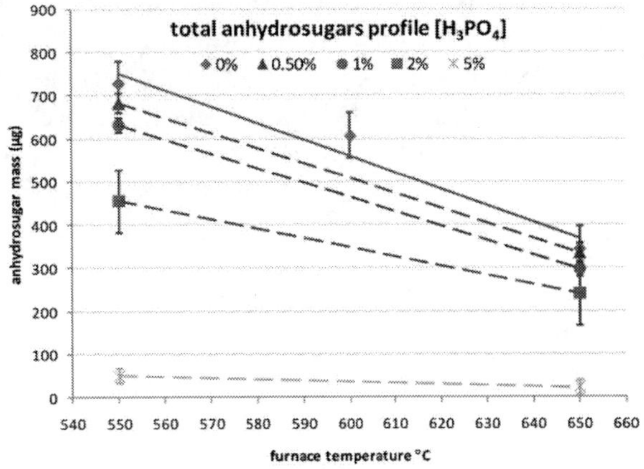

Figure 3b: Anhydrosugar yields from cellulose treated with varying concentrations of H_3PO_4 (0.5wt%, 1wt%, 2wt% and 5wt%)

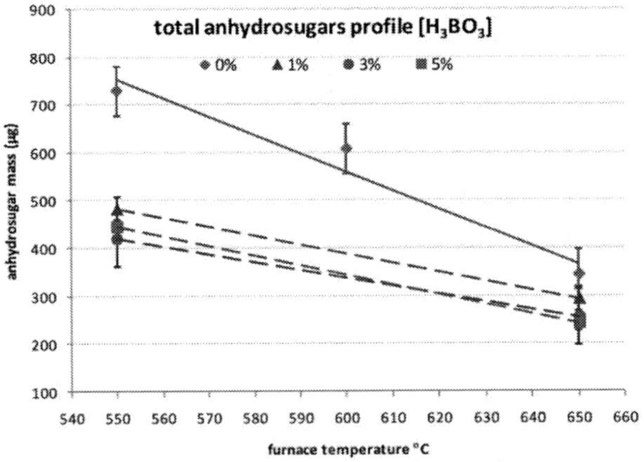

Figure 3c: Anhydrosugar yields from cellulose treated with varying concentrations of H_3BO_3 (1wt%, 3wt% and 5wt%)

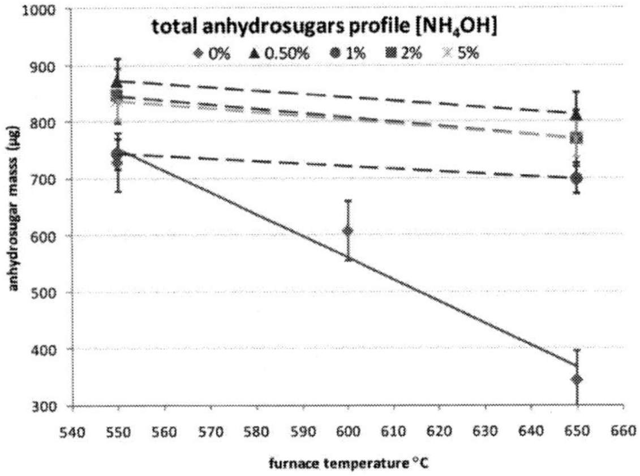

Figure 3d: Anhydrosugar yields from cellulose treated with varying concentrations of NH_4OH (0.5wt%, 1wt%, 2wt% and 5wt%)

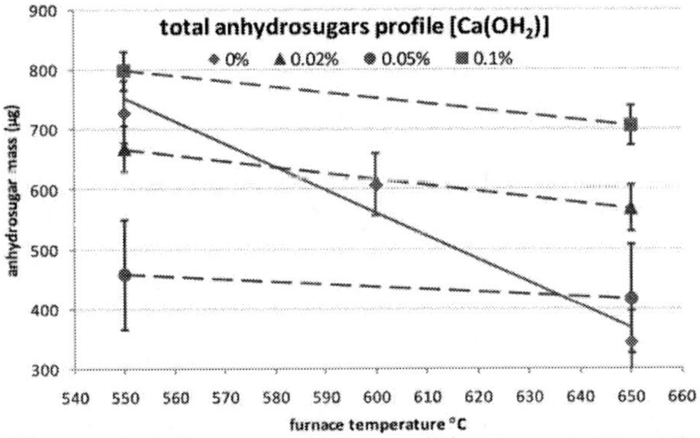

Figure 3e: Anhydrosugar yields from cellulose treated with varying concentrations of $Ca(OH)_2$ (0.02wt%, 0.05wt%, and 0.1wt%)

Looking at the graphs above, we can see that the amount of anhydrosacharides (levoglucosan, levoglucosenone, 1,4:3,6-dianhydro-α-D-glucopyranose, 1,6-anhydro-β-D-glucofuranose) decreases as the reactor temperature is increased. This would be in line with our expectations based on the relative reactions rates for transglycosylation and β-elimination. We can also see that the three acid treatments led to decreased anhydrosacharide yields with higher acid concentrations generally resulting in a corresponding reduction in anhydrosacharides.

The results obtained for cellulose pyrolysis with alkalis were slightly mixed. In the case of ammonium hydroxide, the trend was rather clear that the alkali treatment led to increased levels of anhydrosacharides being produced. However, with the calcium hydroxide treatment, higher levels of anhydrosacharides (compared to untreated cellulose) were only observed at the higher temperatures. In fact, even for ammonium hydroxide, its effects on increasing anhydrosacharides yields were also more pronounced at higher temperatures.

In order to gain further understanding from the results that we have obtained, the relative amounts of levoglucosan, levoglucosenone, 1,4:3,6-dianhydro-α-D-glucopyranose, 1,6-anhydro-β-D-glucofuranose were analysed and is represented in Figure 4.

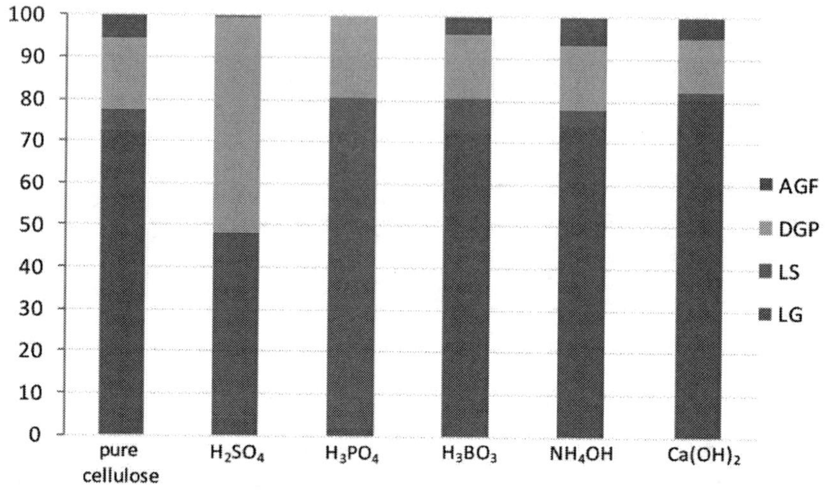

Figure 4: Comparison of the relative yields of the anhydrosacharides [levoglucosan **(LG)**, levoglucosenone **(LS)**, 1,4:3,6-dianhydro-α-D-glucopyranose **(DGP)**, 1,6-anhydro-β-D-glucofuranose **(AGF)**]obtained for the various acid and alkali-treatments.

Figure 4 clearly indicates that treatments with sulphuric and phosphoric acids resulted in higher levoglucosenone levels whereas for boric acid, ammonium hydroxide and calcium hydroxide, the relative amount of levoglucosan was found to be higher.

Based on our results thus far, we can ask the question whether a lower anhydrosacharide yield obtained in the presence of acids was due to the further breakdown of the anhydrosacharides or was it due to promotion of the acid catalysed β-elimination pathway as seen in Figure 1. If we take a closer look at the boric acid treated cellulose runs, the relative amount of levoglucosan was higher compared to the amount of levoglucosenone. This observation can be explained based on the findings of Kawamoto et al. [10] which showed that boric acid forms a stable complex with levoglucosan that prevents further transformation/breakdown. Hence in spite of this, we were still able to observe a decrease in total anhydrosacharide yield for boric acid treated cellulose. This therefore means that the cellulose probably depolymerised via the alternative β-elimination pathway.

On the other hand, we have also seen that at higher temperatures, the alkali-treated cellulose was able to maintain its anhydrosacharide yields better than untreated cellulose. This "stabilising" effect of the alkalis could be due to neutralisation of H^+ ions that are needed to catalyse the β-elimination pathway. This, in turn, allowed the transglycosylation pathway to be favoured during the pyrolysis.

4. Conclusion

Based on these experimental results, it can be seen that the presence of an acid or an alkali can have a significant effect on the yield of anhydrosacharides during pyrolysis. We have observed that acids (H_2SO_4 and H_3PO_4) decreased the total amount of anhydrosacharides and levoglucosan. This is probably due to the acids catalysing the hetrolytic reactions and promoting secondary dehydration reactions which converts levoglucosan into levoglucosenone and 1,4:3,6-dianhydro-α-d-glucopyranose. We have also seen that when the cellulose was treated with alkalis (NH_4OH and $Ca(OH)_2$), the overall yield of anhydrosacharides and levoglucosan increased accordingly.

These two main observations indicate that the β-elimination pyrolysis pathway is likely to be catalysed by the presence of acidic species at the expense of the alternative transglycosylation pathway. This is further confirmed when the presence of an alkali "inhibited" the β-elimination route in favour of transglycosylation.

References

1. **Mamleev V., Bourbigot S., Le Bras M., Yvon J.**, The facts and hypotheses relating to the phenomenological model of cellulose pyrolysis. Interdependence of the steps, *Journal of Analytical and Applied Pyrolysis*, 84, 1-17. 2009.
2. **Graham R.G., Bergougnou M.A.**, Fast pyrolysis of biomass, *Journal of Analytical and Applied Pyrolysis*, 6, 95-135. 1984.
3. **Kawamoto et. al.**, Catalytic pyrolysis of cellulose in sulfolane with some acidic catalysts, *Journal of Wood Science*, 53, 127-133. 2007.
4. **Shen D.K., Gu S.**, The mechanism for thermal decomposition of cellulose and its main products, *Bioresource Technology*, 100, 6496-6504. 2009.
5. **Lin Y.C. et. al.**, Kinetics and mechanism of cellulose pyrolysis, *Journal of Physical Chemistry C*, 113 (46), 20097–20107. 2009.
6. **Budarin et. al.**, The preparation of high grade bio-oils through the controlled, low temperature microwave activation of wheat straw, *Bioresource Technology*, 100, 6064-6068. 2009.
7. **Hassan E.M., Steele P.H.**, Characterization of fast pyrolysis bio-oils produced from pretreated pine wood, *Applied Biochemistry and Biotechnology*, 154, 182-192. 2009.
8. **Banyasz J.L., Li S., Lyons-Hart J., Shafer K.H.**, Gas evolution and the mechanism of cellulose pyrolysis, *Fuel*, 80, 1757-1763. 2001.
9. **Capart R., Khezami L., Burnham A.K.**, Assessment of various kinetic models for the pyrolysis of a microgranular cellulose, *Thermochimica Acta*, 417, 79-89. 2004.
10. **Kawamoto et. al.**, Stable complex formation with boric acid in pyrolysis of levoglucosan in acidic media, *Journal of Analytical and Applied Pyrolysis*, 82, 78-82. 2008.

Organosolv revisited – biomass valorization with new perspectives

Michael Schrems, Philipp Vejdovszky, Martin, Zeilinger, Sonja Bednarik,
Emmerich Haimer, Falk Liebner, Thomas Rosenau, Antje Potthast
Christian-Doppler-Laboratory "Advanced cellulose chemistry and analytics"
Department of Chemistry, University of Natural Resources and Life Sciences,
A-1190 Vienna, Austria
thomas.rosenau@boku.ac.at, antje.potthast@boku.ac.at

ABSTRACT

This study reinvestigated the well known Organosolv concept aiming at maximum lignin output using $scCO_2$ as co-solvent. Lignin is a still underestimated natural resource with potential application in the near future. The use of N-modified lignins as soil improvers was considered being one of the most interesting utilizations in response to progressive desertification in parts of the world. The presented process uses rye straw as a renewable source achieving up to 70% delignification with lignin purities of 80-85% and nitrogen contents of up to 4%. The nitrogen enrichment was achieved by thermal treatment of rye straw under alkaline Organosolv conditions simultaneously using aqueous NH_4OH as nitrogen source and supercritical carbon dioxide as pulping additive. The effectiveness of the used supercritical carbon dioxide could be demonstrated with a model substance study. Apocynol was used as a representative lignin model showing a significantly faster degradation in the presence of $scCO_2$ which is in accordance to a 40% lower Arrhenius activation energy.

I. INTRODUCTION

Due to fading fossil resources, valorization of renewable resources is on the rise.[1] The enormous demand for energy currently covered through petroleum refinery is the main driving force behind this process. The ongoing change to inexhaustible energy sources does not only require new technologies for biomass processing but will also lead to new industry branches for platform chemicals. Bioconversion of renewable resources by treatment with hot aqueous ethanol, known as the Organosolv process, is an old concept dating back to 1931.[2] Although the idea of separating or extracting components out of wooden materials arose already in the early 19th century, the middle of the 20th century can be considered as the time, when this concept drew attention to a larger community. With the beginning of the 1990s the interest in this research area increased remarkable focusing on the delignification of renewable materials, mainly wood. With the establishment as a promising approach for the future, new possible facets appeared in scientific databases. Organosolv pulping as a conceivable „green" alternative for established sulfur-containing processes as kraft and sulfite, generating enormous amounts of non-recyclable by-products, gained widespread interest with its peak in the nineties. In the following years the focus was directed to other possible renewable raw materials for Organosolv processes such as straw and annual plants. The latest development in this area is the use of Organosolv principles as a pretreatment step for various renewable resources to enhance the bioethanol output from polysaccharides in bacterial digestion.[3] Considerable pretreatment effects can also be obtained with simply using hot water but waive delignification to further increase the enzymatic accessibility.[4,5]

Organosolv conditions involve usually a mixture of water and an aliphatic alcohol, mainly ethanol in various mixtures. Comparable results towards yield can be achieved using methanol although higher purity hence delignification of the pulp can be achieved with ethanol.[6] Use of supercritical carbon dioxide as an additional solvent enables further improvements. Super critical carbon dioxide causes an anomalous porosity and lamellar structure within the material resulting in a higher accessibility for the ethanol water mixture.[7,8] The significantly higher solubility of carbon dioxide in ethanol in opposite to water makes the application of an alcohol water mixture even more preferable over pure hot water pretreatment.[9] Pretreatment steps with similar gases such as sulfur dioxide are also capable to enhance the achievable results but have to be considered less applicable because of their inability to reduce the carbon dioxide foot print.[10]

The everlasting issue in energy production based on renewable materials is how to make such a process profitable. Resources like straw are cheap but the energy input for the transformation to biogas is high rendering a simple conversion less feasible. As pulp resulting from Organosolv treatment of straw and other annual plants does not show sufficient quality to enter the paper and fiber industry, other main constituents need to be transformed to value-added products to sustain such a concept. Lignin, one of the main components in all naturally grown materials, shows a high potential to meet this demand although efficient usage could not be established in the past. Until now rather neglected by simply using it as fuel, various more beneficial applications are possible. The usage of Organosolv lignin as a polyphenol source is just one economically interesting alternative, provided pure isolation.

A very promising application is the use of Organosolv lignin as a source of artificial humic substances which can be applied for recultivation of degraded soils or general soil improvement. Progressive desertification in parts of the world renders this product class highly beneficial in the future. Ammonoxidation is one of the technological opportunities for converting lignin and other ligneous biomaterials into nitrogen-rich organo-mineral fertilizers that share a couple of similarities with natural humic substances. The special value of these modified lignins is the slow release of nitrogen due to different binding forms of nitrogen generated upon ammonoxidation. The fact, that substantially delignified renewable materials are in addition much easier converted to degradation products such as bioethanol and biogas, led us to the development of an organosolv-based bioconversion process maximizing the isolated amount of pure lignin. This can be achieved by applying high pressure ideally with carbon dioxide. As the subsequent ammonoxidation is usually executed in aqueous media, a significant amount of energy can be saved by avoiding the intermediate purification of the lignin. Therefore we combined these two process steps in one by performing the N-modification already during the Organosolv process replacing the water by aqueous ammonium hydroxide.

Understanding the delignification process and the effect of supercritical carbon dioxide on a molecular level is considered being crucial for the process design. First investigations towards the process occurring during organosolv pulping were done in the early 2000s by Kishimoto and Sano.[11, 12] In this work further studies were conducted to clarify the effect of super critical carbon dioxide on the delignification process. For this purpose 4-(1-Hydroxy-ethyl)-2-methoxy-phenol (Apocynol) was used as a representative lignin model substance.

II. EXPERIMENTAL

Chemicals

Chemicals were obtained from commercial sources and were of the highest purity available. Rye straw was obtained from different suppliers in Lower Austria and milled to <1mm containing an average water content of 6.3%. Apocynol was obtained via reduction of acetoguaiacone using sodiumborohydride in 52% yield.

Experimental setup

CO_2 assisted Organosolv pulping was performed in a 0.2 *l* autoclave with jacketed heater (SEPAREX, France). 12 *g* of the $scCO_2$-extraction material were mixed with 100 *ml* 50% (v/v) ethanol in water (initial phase) or aqueous ammonium hydroxide (0.1, 2.5, 5.0 and 10.0 w% NH_3, see **Table 1.**) and equilibrated for 15 minutes. After equilibration the mixture was filled into the preheated autoclave (80 *°C*) and pressurized with CO_2 to 6 *MPa*. Subsequently the heating jacket was heated to the desired temperature. After reaching the adjusted value, the system was pressurized using a hand-piston-pump (Sitec, Switzerland) and left under these conditions until the individual reaction time has elapsed. After the treatment the system was depressurized and the solid residue was washed with 400 *ml* 0.1*M* NaOH and subsequently filtrated. The residue was supposed to be enriched cellulose whereas the supernatant was supposed to bear lignin and hemicellulose. In order to separate these fractions, the supernatant was acidified with HCl to a pH lower than 3 leading to precipitation of the high molecular lignin. After filtration the supernatant as well as the two solid fractions were dried at 70 *°C* and subjected to further analysis.

General Analytics

GPC System. For GPC measurements, the system as described by Röhrling et al[5] was used: online degasser, Dionex DG-2410, Kontron 420 pump; pulse damper; autosampler, HP series 1090; column oven, refractive index (RI) detector, Shodex RI-101. DMAc/LiCl (0.9%, w/v), filtered through a 0.02 *μm* filter, was used as the eluant. The sample was injected automatically, chromatographed on two serial GPC columns, and monitored by RI detection. Molecular weight distribution and related polymer relevant parameters were calculated by software programs. The following parameters were used in the GPC measurements: flow, 1.00 *ml/min*; columns, two, PL gel mixed A LS, 20 *μm*, 7.5x300 *mm*; injection volume,: 100 *μl*; run time, 45 *min*. Data evaluation for lignin molecular weight are based on calibration wit narrow polystyrene standards (Astra 5 software). Standards used for calibration: 30 *kg/mol*, 114.2 *kg/mol*, 200 *kg/mol*, 300 *kg/mol*, 700 *kg/mol*.

Curie-point pyrolysis GC/MS was performed using a CPP-40 pyrolyser (GSG) coupled with a GC 6890 and MSD 5973 (AGILENT TECHNOLOGIES). 260 *μg* of the sample was pyrolysed at 600 *°C* (FECRALLOY^{TM}) for 10 seconds. The pyrolysate was carried by helium into the inlet (250 *°C*, split 1:20) of the gas chromatograph. Separation was achieved using a fused silica HP-5ms column (30 *m*, 0.25 *mm*, 25 *μm*) and a column flow of 1.0 $ml\ min^{-1}$. Oven program: 50 *°C* (5 *min*), 5 $°C\ min^{-1}$ to 180 *°C*, than 10 $°C\ min^{-1}$ to 280 *°C* (9 *min*). Auxiliary temperature: 250 *°C* (18 *min*), 10 $°C\ min^{-1}$ to 280 *°C* (14 *min*). Mass spectrometer: EI mode (70 *eV*); ion source 230 *°C*; quadrupole 150 *°C*, $7.6 \cdot 10^{-6}$ *Torr*. Peak assignment was accomplished using the NIST 2005 mass spectra database. Purity evaluation is based on calibration with lignin standards to which 2.5, 5.5 and 10.5 wt.-% of pulp were added. The evaluation of the polysaccharide content of the analysed samples was done by comparing the peaks of furfural and guajacole in the pyrograms. Furfural is a reliable indicator for the sugar content whereas guajacole is the main lignin fragment occurring after thermal fragmentation. Analysis of the hydroxymethyl-furfural peak as main product of hexoses showed that it had only minor effect on the overall result and was therefore not taken into account.

Elemental analysis (C, H, N) was performed at the Microanalytical Laboratory of the University of Vienna.

Analysis of ash and minerals was performed by the Institute of Waste Management at the University of Natural Resources and Life Sciences Vienna. The samples were heated up in aluminium oxide crucibles to 950 *°C* with a rate of 10 *K/min*, gas flow 150 *ml/min* (80% He/20% O_2) using 16mg material per analysis.

HPLC analysis was performed with a Dionex system (P580 pump, GINA 50 autosampler,) running with Chromeleon 4 software. Detector: UVD340U diode detector. Column: Agilent done with a Agilent Eclipse XDB-C18, Injection volume: 10 *μl*, Flow: 0.8 *ml/min*, Gradient program: 0% eluent A→ 95% eluent A (0 *min* → 15 *min*), 95% eluent A (15 *min* → 18 *min*), 0% eluent A (18 *min* → 20 *min*), Detection wavelength: 280 *nm*; Eluent A: 100% MeCN; eluent B: 100% Water.

Statistics

Statistical experiment design was done with CADEMO (Computer Aided Design of Experiments and Modeling) from BioMath GmbH using a Box-Behnken design. Design and calculation were done by Department of Applied Statistics at the University of Natural resources and Life Sciences Vienna.

Synthesis of apocynol

To a solution of acetoguaiacone (7 *g*, 42 *mmol*) in 212 *ml* 56% ethanol 3.5 *g* (93 *mmol*) of $NaBH_4$ are added in portions and afterwards stirred for 10 *h* at room temperature. Another portion of $NaBH_4$ (1 *g*, 26 *mmol*) is added and stirring is continued for additional 14 *h*. The reaction mixture is neutralized with CO_2 (dry ice!), introduced through a glass capillary with a wide diameter to prevent plugging, diluted with 200 *ml* water and extracted six times with 150 *ml* Et_2O. Combined organic phases were dried over Na_2SO_4 (about 6 *h*) and constricted to 200 *ml* under reduced pressure. The ethereal residue is transferred into a crystallization dish and recrystallized when all solvent evaporated. This step is necessary to prevent formation of dimers. The crystallized product is solved in small amounts of ethyl acetate (the unsolvable dimers are filtered off) and precipitated with hexane. The obtained white crystals are dried under reduced pressure. The remaining solution can be constricted and the yellowish crystals can be recrystallized again in the same matter to higher the yield (3,7 *g*; 52%).

^{13}C NMR (75 MHz, $CDCl_3$): 146.61 (*C*-3), 144.98 (*C*-4), 137.93 (*C*-1), 118.32 (*C*-6), 114.15 (*C*-5), 107.99 (*C*-2), 70.33 (*C*-α), 56.00 (OMe), 25.09 (*C*-β)

III. RESULTS AND DISCUSSION

The first part of this study will present investigations towards the Organosolv process. In the initial phase various chemicals e.g. acids and bases have been tested in different concentrations as additives to potentially raise the lignin output and or the purity. Alterations in the consumption of the washing solutions were conducted for the same purpose. After this phase a statistical approach was used to find optimal values for the process parameters. The second part will reveal new insights into the delignification process on a molecular level. Experiments have been done under high pressure at various temperatures. The influence of super critical carbon dioxide was investigated to understand this enhancing factor in more detail.

III.1 Development of an Organosolv process aiming at maximal lignin output

The Organosolv pulping process is known to work best with equivalent alcohol water ratios at temperatures at or slightly above 180 *°C* when high lignin output is desired. The reaction time appears to be directly proportional to the yield leading to increasing yields with increasing time.[3] With this knowledge the test parameters for the initial phase were chosen to be 180 °*C*, 16 *MPa* and 160 *min*. The rather high pressure is due to several factors. Carbon dioxide is known to change into the supercritical state about 31 °*C* and 7.3 *MPa*. More important is the requirement to keep water and also ethanol in the liquid state at this temperature. Additionally high pressures guarantee a high concentration of carbon dioxide in the ethanol-water mixture.

The first tests showed that ethanol gives only slightly better yields than methanol with comparable purity of the obtained lignin fraction. As methanol is approximately twenty percent cheaper than ethanol we continued the study with methanol as the co-solvent. Addition of acid (H_2SO_4) gave good yields for the lignin but very low yields for the cellulose fraction. As this fraction is also relevant for the economics of this concept, this additive has to be considered as unfavorable. The addition of base (NaOH) turned out to be similar unfavorable as this caused very impure lignin fractions. The effectivity of carbon dioxide was proven with two experiments under atmospheric pressure and at 160bar using nitrogen instead of $scCO_2$. The obtained yields were 38% and 13%

lower, respectively. These results indicate that the high pressure itself causes increasing yields but only carbon dioxide can raise the yield even further. Experiments with the washing solution showed that it is not possible to extract lignin with pure water. Very good results could be achieved with 1 *N* NaOH but this led to very high salt concentrations and impurities in the lignin fractions. The usage of 0.1 *N* NaOH gave just 6% lower yields but the salinity and the purity were significantly better. 1 *N* NH_4OH also showed some potential as washing solution but gave 60% lower yields rendering this alternative not very suitable. The determination of the lignin molecular weight M_w was done with GPC analysis. Due to the strong fluorescence of the samples it was not possible to determine the molecular weight with light scattering. Therefore the column was calibrated with polystyrene standards ranging from 30-700 *kg/mol*. The obtained M_w values from the analyzed lignin fractions were relatively homogenous with an average of 7.9 *kg/mol* ± 1.3. Analysis of the ash and mineral content displayed a rather low concentration within the lignin fraction with average ash content of 3.1 % ± 1.5. The next step was to find the optimal operating parameters for the described process. For this purpose a statistical approach as used to reduce the overall number of experiments and avoid systematic errors. The limits for an effective process were considered to be between 150-180 *°C* with pressures between 60-160 *bar*, reactions times between 10-160 *min* and ammonium hydroxide concentrations between 0.1 and 10%. **Table 1** gives an overview about these experiments including achieved yields and purity for the cellulose and the lignin fraction. All purity values have been determined by Curie-point pyrolysis GC/MS.

The bold black highlighted tests 8 and 22 in **Table 1** turned out to be the best pulping conditions when aiming for maximal lignin output. With a straw lignin content of about 20% both conditions lead to delignification of approximately 70% with a lignin purity of 80-85%. These results also show that there is a strong dependency towards all reaction parameters. Depending on the situation it can be chosen between a shorter but higher energy inputs or longer less intensive conditions. With average nitrogen enrichment of 4% in the lignin fraction this process can be considered as a promising concept for the production of nitrogen-rich lignins the latter being worth to become subjects of a thorough study evaluating their fertilizing and soil improving effects as well as their bio- and phytoactivity.

Table 1: *Statistical experiment design after Box-Behnken*

	T [°C]	p [bar]	t [min]	NH_4OH conc. [%]	Residual cellulosic material [%]	Polysaccharide content pulp [%]	Obtained Lignin [%]	Polysaccharide content lignin [%]
	Block points							
1	157	85	47	2.5	72.3	60.9	1.8	4.7
2	172	85	47	2.5	68.8	60.8	7.7	*n.a.*
3	157	135	47	2.5	75.0	57.5	8.0	*n.a.*
4	172	135	47	2.5	66.1	57.6	4.5	9.0
5	157	85	122	2.5	67.9	55.7	12.5	*n.a.*
6	172	85	122	2.5	59.8	56.7	4.5	12.9
7	157	135	122	2.5	70.5	53.7	2.7	*n.a.*
8	**172**	**135**	**122**	**2.5**	**65.2**	**55.8**	**14.3**	**19.7**
9	157	85	47	7.5	77.7	64.9	8.9	15.8
10	172	85	47	7.5	67.9	64.6	5.4	10.3
11	157	135	47	7.5	59.8	62.4	2.7	8.0
12	172	135	47	7.5	71.4	67.5	11.6	20.3
13	157	85	122	7.5	66.1	72.3	3.6	9.4
14	172	85	122	7.5	72.3	61.3	10.7	11.1
15	157	135	122	7.5	74.1	63.6	9.8	*n.a.*
16	172	135	122	7.5	52.7	*n.a.*	5.4	10.8
	Face points							
17	150	110	85	5.0	62.2	61.6	6.3	9.8
18	180	110	85	5.0	58.9	56.7	7.1	6.8
19	165	60	85	5.0	63.4	57.3	11.6	9.8
20	165	160	85	5.0	59.8	62.0	7.1	*n.a.*
21	165	110	10	5.0	74.1	52.3	3.6	*n.a.*
22	**165**	**110**	**160**	**5.0**	**86.6**	**68.0**	**14.3**	**14.2**
23	165	110	85	0.1	64.3	*n.a.*	8.0	*n.a.*
24	165	110	85	10.0	67.9	55.8	7.1	*n.a.*
	Center point							
25	165	110	85	5.0	67.0	67.6	8.9	*n.a.*

III.2 Investigation of the reaction kinetics through model substance studies

The aim of this study was to obtain a deeper understanding of the reactions occurring during the Organosolv pulping process in the presence of carbon dioxide. The focus was placed on the delignification process in dependence on the temperature applied and the role of supercritical carbon dioxide. For this purpose apocynol seemed to be a suitable model substance for the investigation of the delignification process. It is easily accessible through reduction of acetoguaiacone and representative due to its structure similar to lignin monomer units. [13]

The experiments were conducted in a metal reactor built for high pressure experiments which was pre-heated to 80 *°C*. 20 *mg* of apocynol (**1**) was placed in a small beaker and dissolved in 10 *ml* of a 1:1 mixture of ethanol and water. The beaker with the solution was placed in the reactor and the reaction chamber sealed afterwards. An initial CO_2 pressure of 6 *MPa* was applied to ensure just liquid phases already before reaching the adjusted temperature. The end temperature (140-180 *°C*) was reached within several minutes followed by adjusting a pressure of 16 *MPa* (t=0). At the end of the individual reaction time the heat control was turned off and the metal sheath was cooled with ice to accelerate cooling. Eventually, the orange coloured liquid obtained was analysed by HPLC.

Experiments conducted at 140 *°C*, 160 *°C* and 180 *°C* with and without supercritical carbon dioxide showed that there is an increase in degradation of the model substance with increasing temperature (cf. **Fig. 1**). But this increase is only of minor relevance compared with a much stronger degradation effect in presence of super critical carbon dioxide. As indicated in figure 1, the degradation rate at 140 *°C* with $scCO_2$ present is even faster than the reaction rate at 180 *°C* without $scCO_2$. The order of this reaction was estimated as first order. The calculation of the Arrhenius activation energy gave a significantly lower activation energy for this reaction when $scCO_2$ was applied. An E_A of 68.9 *kJ/mol* for the reaction without $scCO_2$ in comparison to 42.0 *kJ/mol* points out that the usage of $scCO_2$ can lower the activation energy of approx. 40%.

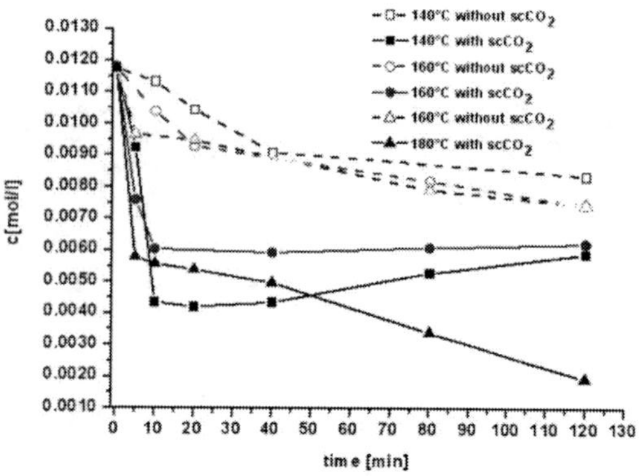

Fig. 1: Degradation of apocynol at different temperatures with and without $scCO_2$

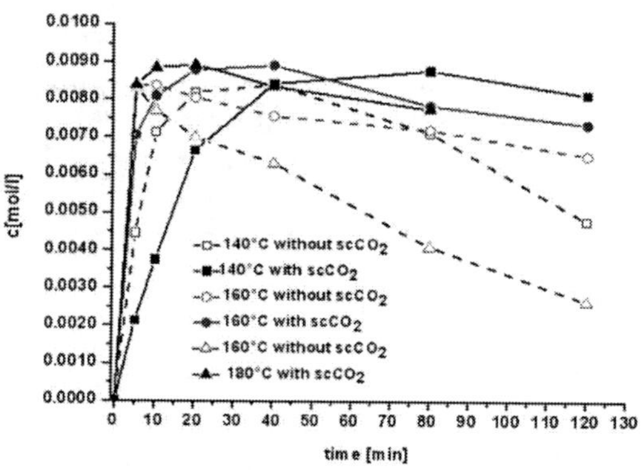

Fig. 2: Formation and transformation of the reaction intermediate 2-methoxy-4-vinylphenol

HPLC analysis of the samples revealed that the overall degradation reaction of the model substance apocynol (**1**) is a combination of two consecutive reactions. The first step is the degradation of the apocynol is the transformation into 2-methoxy-4-vinylphenol (**2**), formed via a *para*-chinone methide (*p*QM), which stabilizes into (**2**) by a [1,7]-sigmatropic proton shift.[14] The second step is the formation of 4-(1-ethoxyethyl)-2-methoxy-phenol (3) by addition of ethanol to the vinyl intermediate according to a pseudo-first-order rate law, cf. **Fig. 3**.

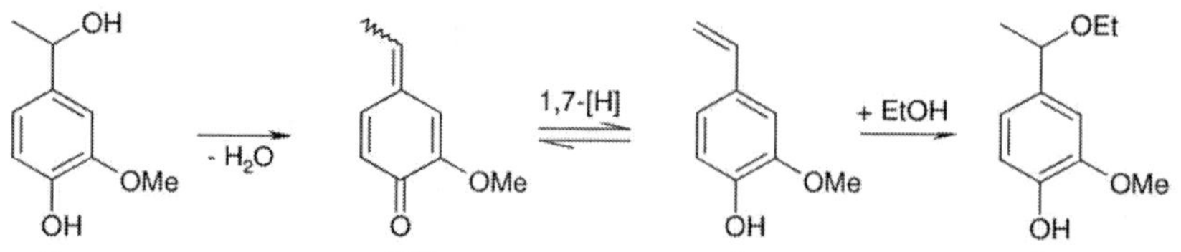

1 ***p*QM** **2** **3**

Fig. 3: Reaction pathway of apocynol under Organosolv pulping conditions in aqueous ethanol (50%, v/v)

The presence of $scCO_2$ slows down the subsequent transformation of the vinyl intermediate (**2**) into 4-(1-ethoxyethyl)-2-methoxy-phenol (**3**), cf. **Fig. 2**. Whereas in the absence of CO_2 the intermediate (**2**) is transformed within minutes into the final product (**3**), the presence of $scCO_2$ causes a very fast formation of the stable intermediate (**2**) but a decelerated transformation into the end product. Even the effect of temperature almost vanished. The presence of $scCO_2$ also effects the zero order formation of the end product 4-(1-ethoxyethyl)-2-methoxy-phenol (**3**). Although the final concentration is higher when $scCO_2$ is applied the initial reaction rate is definitely higher when no $scCO_2$ is present. This finding is in accordance with the calculated values for the Arrhenius activation energy. The E_A for this reaction without $scCO_2$ was determined with 4.6 *J/mol* which is about four to five times lower than the calculated value of 20.9 *J/mol* for $scCO_2$.

IV. CONCLUSIONS

The developed process allows isolating major parts of the total lignin content in rye straw. Significant increase of the lignin output can be achieved by applying high pressure during the process. Highest amounts can be obtained when $scCO_2$ is used during the reaction. This effect can be explained by the super critical state of the carbon dioxide at the investigated reaction conditions making the raw material better accessible and acting as a solvent enhancer. A statistical approach was used to find the optimal regions for all operating parameters. Temperature, pressure and time had a significant influence on the results. Additional tests were done to evaluate the contributions of various additives and auxiliaries namely acids and bases and the possible exchange of ethanol for methanol which can save up to twenty percent in the overall solvents costs.
The study showed that it is possible to achieve a delignification of approximately 70% with a sufficiently pure lignin fraction. Furthermore it was proven that nitrogen enrichment, conventionally performed subsequently, can be accomplished during the pulping process significantly reducing energy demand. A nitrogen enrichment of 4% within the lignin fraction is considered to be sufficient for the aimed purpose of soil improvement.

Kinetic studies clearly showed the influence of supercritical carbon dioxide on the reaction mechanism. Apocynol was used as a simple model for lignin. Degradation kinetics with and without super critical carbon dioxide were recorded at different temperatures between five minutes and two hours. The decay of the original model substance and the formation of degradation products were analyzed by HPLC. The use of super critical carbon dioxide during the process does not only accelerate the decay of the model substance but also lowers the activation energy distinctly, which is in accordance with current literature. It was shown that the presence of $scCO_2$ only accelerated the transformation into the intermediate product (**2**) but slowed down the formation of the end product from that intermediate. It seems quite clear that the application of super critical carbon dioxide has a positive impact on the delignification of the designed process. Tests replacing the $scCO_2$ with nitrogen within the process and the accelerated transformation of the model compound apocynol confirmed the expectations of a supportive effect. The hindrance of $scCO_2$ on the formation of the apparently end product 4-(1-ethoxyethyl)-2-methoxy-phenol can be explained with a potentially different reaction environment making the intermediate compound more stable. To get a better insight in the role of $scCO_2$ on the delignification process, a study using a dimeric compound containing a β-O-4 ether bond similar to the one used in experiments of Kishimoto and Sano is planned.[11]

V. ACKNOWLEDGEMENT

The financial support by the Christian Doppler Research Society (CD-laboratory "Advanced Cellulose Chemistry and Analytics") and Blue Globe Energy, Mannheim, Germany, is gratefully acknowledged. The authors would like to thank Prof. Dr. Ena Smidt from the Institute of Waste Management at the University of Natural Resources and Life Sciences Vienna for performing the ash and minerals content analyse and Prof. Dr. Karl Moder from the Department of Applied Statistics at the University of Natural resources and Life Sciences Vienna for the statistical experimental design.

VI. REFERENCES

1. J. H. Clark, V. Budarin, F. E. I. Deswarte, A. J. Wilson, Green chemistry and the biorefinery: a partnership for a sustainable future, *Green Chem.* **2006**, *8*, 853-860.
2. T.. Kleinert, K. Tayenthal, Ueber neuere Versuche zur Trennung von Cellulose und Inkrusten verschiedener Hölzer, *Angewandte Chemie* **1931**, *44*, 788-791.
3. X. Pan, N. Gilkes, J. Kadla, K. Pye, S. Saka, D. Gregg, K. Ehara, D. Xie, D. Lam, J. Saddler, Bioconversion of Hybrid Poplar to Ethanol and Co-Products Using an Organosolv Fractionation Process: optimization of Process Yields, *Biotechnology and Bioengineering* **2006**, *94(5)*, 851-861.

4. M. Zeng, N. S. Mosier, C-P. Huang, D. M. Sherman, M. R. Ladisch, Microscopic Examination of Changes of Plant Cell Structure in Corn Stover Due to Hot Water Pretreatment and Enzymatic Hydrolysis, *Biotechnology and Bioengineering* **2007**, *97(2)*, 265-278.

5. K. Öhgren, R. Bura, J. Saddler, G. Zacchini, Effect of hemicellulose and lignin removal on enzymatic hydrolysis of steam pretreated corn stover, *Bioresource Technology* **2007**, *98*, 2503-2510.

6. M. Oliet, J. Garcia, F. Rodriguez, M. A. Gilarranz, , Solvent effects in autocatalyzed alcohol-water pulping – Comparative study between ethanol and methanol as delignifying agents, *Chem. Eng. J.* **2002**, *87*, 157-162.

7. M. Gao, F. Xu, S. Li, X. Ji, S. Chen, D. Zhang, Effect of SC-CO_2 pretreatment in increasing rice straw biomass conversion, *Biosystems Engineering* **2010**, *106*, 470-475.

8. D. Pasquini, M. T. B. Pimeta, L. H. Ferreira, Sugar cane bagasse pulping using supercritical CO_2 associated with co-solvent 1-butanol/water, Journal of Supercritical Fluids **2005**, *34*, 125-131.

9. E. Dalmolin, E. Skovroinski, A. Biasi, M. L. Corazza, C. Dariva, O. Olivieira, Solubility of carbon dioxide in binary and ternary mixtures with ethanol and water, *Fluid Phase Equilibria* **2006**, *245*, 193-200.

10. P. Sassner, M. Galbe, G. Zacchi, Steam Pretreatment of *Salix* with and without SO_2 Impregnation for Production of Bioethanol, *Applied Biochemistry and Biotechnology* **2005**, *121-124*, 1101-1118.

11. T. Kishimoto, Y. Sano, Delignification Mechanism during High-Boiling Solvent Pulping. Part 1. Reaction of Guaiacylglycerol Ether, *Holzforschung* **2001**, *55*, 611-616.

12. T. Kishimoto, Y. Sano, Delignification Mechanism during High-Boiling Solvent Pulping. Part 2. Homolysis of Guaiacylglycerol-β-Guaiacyl Ether, *Holzforschung* **2002**, *56*, 623-631.

13. E. Adler, S. Delin, G. E. Miksche, Synthese des Isoeugenolglykol-β-(2-methoxyphenyl)-äthers, *Acta Chem. Scand.* **1966**, 20(4), 1035-1037.

14. D. Fengel, G. Wegener, Wood: chemistry, ultra structure, reactions, deGruyter, Berlin, New York, **1989**.

This welcoming session sponsored by Fuels & Petrochemicals Division (F&PD) is a new initiative organized by the AIChE's national and F&PD technical programming team as a preamble to the division's sponsored technical sessions. This will be opened by a short introduction by the chair of the division. Besides recognizing leadership in his introductory short presentation, the chair will highlight the key elements of value additions the division brings to its members. The presentation will be concluded with a special request to join the division where professional opportunities are limited only by members' imagination, to positively impact academia and industry and thereby society at large. For details of the division and its many facets of activities and functions please follow the link www.aiche-fpd.org.

(10b) What Are We Going to Do with All This Natural Gas? - Ramifications of the Dramatic Expansion of Natural Gas Available to the Market Due to North American Shale Gas Production and LNG Liquefaction

Sam Thomas, Chart Industries, Inc., Garfield Heights, OH

The success in Shale and other unconventional formations has dramatically expanded the N. American Natural Gas reserves and promises to also do so in many other parts of the world. New LNG liquefaction capacity in the Middle East and Asia along with additional large scale capacity from Australia and Papua New Guinea coming on-line in the next 5 years opens the opportunity for a global trade in Natural Gas stabilizing historically volatile pricing. How will this important energy and petrochemical feedstock be utilized? The success in improving techniques to allow unconventional gas reserves to be proven and significantly increase the available reserves has not as yet been fully appreciated by markets and policy makers. This presentation reviews the ramifications of natural gas becoming a larger part of the global energy supply for which there are numerous competing interests. Natural Gas holds the ability to significantly reduce carbon emissions by substitution for coal in power generation, it can be used as a transportation fuel either as LNG, CNG or GTL clean diesel increasing energy independence of some of the largest economies in the world simultaneously reducing carbon emissions. It could enable the USA to maintain its position as a leading petrochemical producer if used for ethylene feedstock. How will these competing interests play out and how will governments respond?

See more of this Session: Fuels & Petrochemicals Division Welcome Session and Energy Leadership Presentation
See more of this Group/Topical: Topical 7: 14th Topical on Refinery Processing

(14a) The Energy Challenge: Providing Sustainable Supplies to Meet Growing Demand

F. Emil Jacobs, ExxonMobil Research and Engineering Company, Annandale, NJ

As economies grow and living standards improve, the world faces a dual energy challenge: meeting the growing demand for energy while mitigating emissions and their impact on the environment. ExxonMobil believes that addressing this dual challenge requires an integrated set of solutions which includes increasing the supply of all commercially viable energy sources, reducing emissions, and improving energy efficiency in all sectors of the economy. Technology will be a key enabler of continued progress in all of these areas.

Many experts agree that the immense scale of global energy demand necessitates a continuing role for oil and gas as a substantial portion of the energy supply mix for at least the next several decades. These experts also forecast that the overall energy mix will shift toward a relatively higher level of sustainability through a combination of increased efficiency across the economy, growth in lower-carbon sources of energy, and development and deployment of technologies that help mitigate emissions.

ExxonMobil has developed a number of technologies that contribute to increased supplies of energy, reduced emissions, and higher efficiency in its operations and in consumer use of its products. The presentation will provide an overview of ongoing research and development activity in a broad range of technology areas including algae-based biofuels. It will also highlight some of company's current technologies including lithium-ion battery separator films, lightweight plastics that improve automotive efficiency, and processes that could advance carbon capture and storage.

See more of this Session: Spring Meeting Keynote Address - Sponsored by the AIChE Fuels & Petrochemicals Division
See more of this Group/Topical: Liaison Functions

COMMERCIAL SCALE TRITIUM RECOVERY FROM EXIT SIGNS AND OTHER ILLUMINATION DEVICES AT LLNL

Diane K. Spencer, J. Mark Mintz, Matthew G. Lucas and Dale M. Holck

LLNL-CONF-467413

Lawrence Livermore National Laboratory
Livermore, CA 94551
spencer2@llnl.gov, mintz1@llnl.gov, lucas3@llnl.gov

Abstract

The disposal of exit signs and similar commercially available tritium-powered illumination devices in landfills can pose potential public health and environment risks. These risks can significantly be reduced by collecting the gaseous tritium remaining in these devices. Since 2001, Lawrence Livermore National Laboratory (LLNL) and the U.S. Army have been engaged in a successful tritium recovery and recycle program for a limited number of specialized devices. This laboratory-scale work has been accomplished by hand with reasonable efficiency. However, recent requests for assistance with similar work include millions of these devices and would require much larger capabilities. Thus, LLNL has installed a commercial-scale tritium recovery system—the Tritium Grinder System (TGS)—in its Tritium Facility to complement its laboratory-scale tritium recovery capabilities. The TGS was built to efficiently process tritium-bearing illumination devices on a large scale. It can process up to 72 standard exit signs in a batch, with significantly more for smaller illumination devices (e.g., 2250 rotary phone dials per batch). The TGS liberates gaseous tritium from these various devices by mechanically grinding (or shredding) them under negative pressure in a modified industrial shredder. The liberated tritium is collected in a storage tank and periodically removed by trapping on a regenerable hydrogen getter. This paper presents the results from initial grinder operations, including the collected gas composition and unit processing rate, along with a summary of the design features, operating history, and safety controls.

Overview

Studies of active landfills show leachate with tritium considerably above background, with the tritium concentration in many above the Environmental Protection Agency drinking water MCL of 20,000 pCi/L (Ref. 1). Thus, the Nuclear Regulatory Commission (NRC) regulates tritium-containing illumination devices in general and exit signs in particular requiring reporting and handling requirements, including proper disposal of unwanted or unused signs. The requirements in most cases prevent disposal as ordinary trash. LLNL has recognized that the residual tritium in these discarded devices is valuable and has developed a method to capture it.

TGS Design and Process

The "grinder" portion of the TGS is a commercially available shredder unit from SSI Shredding Systems, Inc., modified to operate under negative pressure. A batch of devices is delivered to the grinder by a conveyor belt and is mechanically shredded. The released tritium, along with decay product helium-3 (3He), and argon (Ar) added as a sweep gas, is then pumped to a collection tank. The collected gas is processed through a commercially available getter to remove the tritium. Solid debris (e.g., exit sign housing residuals) generated through the recovery process is collected in a lined 55-gallon drum using a "bagout" process engineered to minimize outgassing and worker exposure. Upon completion of the bagout process, the content of the drum is disposed of as low-level radioactive waste.

Loading, Grinding, and Tritium Collection

Figure 1 shows the major components of the grinding and collection portion of the TGS. Illumination devices to be processed are introduced through the access door and loaded onto the conveyer belt. Small devices (e.g., telephone dials) are placed inside baskets. After the conveyor is loaded, the access door is closed and the system is pumped down. Argon purging is employed to minimize residual air and water vapor. When the system reaches a suitable negative pressure, the illumination devices and baskets are shredded, with the debris accumulating in the discharge hopper. Strategically positioned cameras allow operators to follow the progress of the baskets up the conveyor and into the grinding chamber. Evolved gas is pumped through a particulate filter to the collection tank. A series of purges with argon is then used to ensure a more complete transfer of the tritium gas from the grinder system.

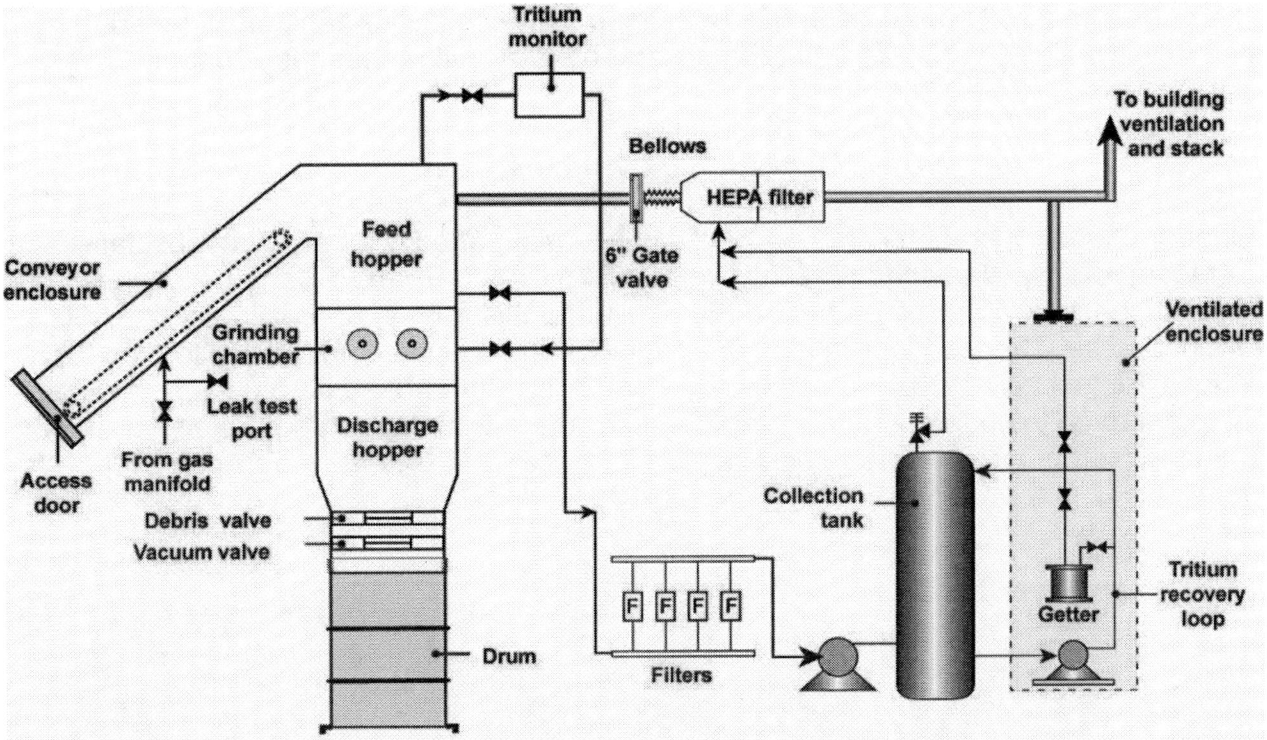

Figure 1. Major components of the TGS.

Following the final argon purge and pump out, the grinder enclosure is let up to ambient pressure, then exhausted to the building ventilation to permit tritium sampling, disposal of debris, and conveyor reloading. A dedicated tritium monitor is consulted prior to venting to verify that less than 1 Ci of tritium is released to the building ventilation when the grinder system is opened. A high efficiency particulate air (HEPA) filter protects the building ventilation from possible particulate formed during the shredding process and, in turn, the building ventilation provides airflow to protect workers from residual contamination.

The grinder is equipped with a double valve near its base. The vacuum valve is relied upon to contain tritium during processing. The debris valve above the vacuum valve supports the accumulated shredded material.

The solid debris generated during the recovery process is safely collected in a lined 55-gallon drum using a "bagout" technique similar to that developed for actinide gloveboxes. This technique allows for the waste drum liners to be sealed and replaced without external exposure. Periodically, a video camera and a small vacuum cleaner nozzle are inserted through a small slit in the liner to inspect and clean the vital vacuum valve seal. The slit is then tape sealed when the camera is removed. Following closure of the drum liner, the filled low-level waste drum is set aside for disposition, and new drum is set in its place.

Tritium Purification, Storage, and Disposition

The central feature of the tritium recovery loop is a high-capacity tritium getter (SAES CapaciTorr-B, MK5 Type). The recovery loop is used to remove tritium (and other active gases) when the pressure in the collection tank reaches 700 torr or 600 Ci, whichever is reached first. The relatively pure argon gas remaining in the collection tank after gettering is exhausted to the building ventilation, but the possibility of recycling that gas is under consideration.

Operating Experience

Grinder startup operations are conducted in two trial period phases using obsolete British Telecom rotary dial telephones used from 1964 to 1981. Phase 1 began on August 11, 2010, and concluded on September 15, 2010. During this phase, approximately 17,250 dials were processed. Phase 2 operations are ongoing, and include collecting tritium on the getter. Phase 2 is expected to be completed by April 2011. Both phases are designed to gain sufficient experience to understand the hazards associated with grinding operations and develop and implement the required controls.

Recovered Tritium

The dials processed consist of a tritium-filled, horseshoe-shaped glass tube glued onto an 84-mm diameter ABS plastic backing plate. The United Kingdom estimated that each dial contains approximately 40 millicuries (mCi) of tritium, with approximately 5 mCi as tritium gas and 35 mCi as tritium embedded in the solid matrix; the latter is not tightly bound. Data collected during Phase 1 operations showed that of the 550 Ci of tritium (HT, DT or T_2) associated with the 17,250 phone dials processed, 58 Ci were recovered. This 10.5% recovery rate compares well with the 12.5% expected from earlier measurements in the United Kingdom. The slight discrepancy may be the result of tritium labeled as methane and water known to be present in trace amounts, but not yet accounted for quantitatively.

Production Rate

Based on results to date, TGS operators can comfortably process two batches of four baskets each of low-curie illumination devices such as phone dials in an average "production" day. Operations involving higher curie items such as signs (~5 Ci each) would likely be limited to one batch a day because of a facility requirement to remove tritium from the collection tank when it approaches 600 Ci.

Collected Gas Analysis

The collection tank pressure was 320 torr at the end of Phase 1. Table 1 shows a mass spectrometric analysis of the collected gas; Table 2, the combined mole fractions of the hydrogen species. The vast majority (99%) is the purged gas Ar, followed by air constituents (O_2, N_2, and CO_2) totaling 0.86%. The presence of air is not surprising, though the O_2/N_2 fraction at 4.5, roughly the inverse of air, is interesting and could be due merely to differences in the Ar sweep efficiency.

Turning to the light isotopes, the preponderance of protium (^1H) at 93% of total hydrogen is likely the primary result of tritium exchange with environmental protium present as water vapor in the glass walls of the phone dials tritium-containment tubes (Ref. 2). Protium originating as molecular hydrogen (H_2) within the glass also may contribute to this value. The mole fraction of protium is estimated by adding the mole fractions of H_2 and half the mole fractions of hydrogen tritide (HT) and hydrogen deuteride (HD), yielding 0.0693 mole fraction %. The total hydrogen species is 0.074 mole fraction % (see Table 2).

The presence of deuterium (D2, DT, HD) can be explained as a known impurity in the original fill gas. The amount of deuterium collected is broadly consistent with the known history of these dials. For example, the approximately identical mole fractions of tritium and deuterium are reasonable given that more than 3 half lives of decay ('average' dial filled in 1972) have converted most of the original tritium fill to ^3He. Tritium half life is 12.5 years and decays to ^3He. After 38 years of decay, approximately 12% of the tritium remains. This suggests that an original deuterium fraction on this order is reasonable and well within the significant uncertainties involved in this estimate.

Table 1. Collected Gas Composition.

Compound	Mole Fraction (%)
Ar	99
O2	0.67
N2	0.15
CO2	0.033
H2	0.066
^{3}He	0.026
He	0.0085
HT	0.0035
HD	0.0032
D2	0.00078
T2	0.00045
DT	0.00036
Methanes	trace
Waters	trace
^{38}Ar	trace
Total	**100**

Table 2. Hydrogen Species Mole Fractions.

Compound	Mole Fraction (%)
Hydrogen (H2, HD/2, HT/2)	0.069
Deuterium (D2, HD/2, DT/2)	0.0026
Tritium (T2, HT/2, DT/2)	0.0024
Total Hydrogen Species	**0.074**

The observed ^3He/tritium ratio of 11 is more difficult to interpret quantitatively due to hydrogen/helium permeation rate differences through the phone dial glass containment tubes and other factors. In the absence of permeation or other differential loss mechanisms, however, a $^3He/T_2$ ratio of about 15 would be observed. This result suggests that permeation of ^3He is faster than tritium through the dial glass as would be expected (Ref. 3).

Hazards and Controls

Phase 1 grinder operations were designed primarily to gather information on the nature and type of radiological hazard the recovery process might pose. Very conservative assumptions were made about the nature and characteristics of tritium contamination that might be generated by the grinding process. In particular, the possibility that residual fine particulates incorporating with or otherwise being contaminated with tritium (i.e., Special Tritium Compound) (Ref. 4) could be a hazard to workers was taken into account. Thus, the area where grinder operations are performed is treated as an "airborne contamination area" and a very high degree of worker protection is specified. Worker protection measures include use of hooded Powered Air Personal Respirators (PAPR, 3M Model GVP-100) and tape-sealed gloves and booties, along with extensive airborne particulate and surface swipe monitoring programs.

Tritiated Particulates

Breathing zone air samplers and continuous passive air samplers ("goose neck" monitors) confirmed that airborne particulates were either extremely low in concentration or not present and therefore not a problem for grinder operations with phone dials. The breathing zone samplers have been discontinued for subsequent operations, but the continuous passive air samplers are being used in the Phase 2 trial period. Thus, the posting in the highest work area was downgraded from "Airborne Contamination Area" to "Contamination Area."

Spreadable Contamination

Consistent with past experience with manual processing of U.S. Army illumination devices, moderate levels of surface contamination (2000–5000 dpm/100 cm^2) were found in several working areas. Light decontamination (mopping) between grinding sessions has proven effective in keeping this spreadable contamination under control.

Personnel Protective Equipment

For Phase 2, use of a full-face HEPA respirator is required only when the conveyor access door is open or while performing bagout-related activities.

Path Forward

The operational trial period for TGS is expected to end April 2011. After that, exit signs and other illuminating devices as well as telephone dials will be processed routinely. Helium-3 is a valuable isotope for fusion research and other applications; the facility is exploring the feasibility of also capturing this material from obsolete illumination devices.

Acknowledgments

This work performed under the auspices of the U.S. Department of Energy by Lawrence Livermore National Laboratory under Contract DE-AC52-07NA27344.

References

1. *Lack of Tritium Exit Sign Control and Contamination of Landfill Leachate*, Association of State and Territorial Solid Waste Management Officials; ASTSWMO Radiation Focus Group, Federal Facilities Research Center, Washington, DC, July 2009.
2. *Tritium in Scottish Landfill Sites*, Report prepared for the Scottish environment Protection Agency by Galson Sciences Ltd., May 2000.
3. Shelby, J. E., Handbook of Gas Diffusion in Solids and Melts, ASM International, Materials Park, OH, April 1996.
4. DOE-HDBK-1129-2008, Tritium Handling and Safe Storage, U.S. Department of Energy, Washington, DC., December 2008.

(21b) Behavior Physics-Mechanical Properties of the Cementitious Matrix From Arum,, Vegetal Fiber, Cement, Metacaulnita and Sand

Goreti M. S. M.[1], **Felix Gamarra**[2], Ramildo D. ToliÍdo filho[1] and **Raimundo Vasconcelos**[3], (1)Civil Engineering, Federal University of Rio de Janeiro, Rio Janeiro, Brazil, (2)Department of Gama, University of Brasilia, UnB Gama College, Brasilia, Brazil, (3)Department of Technology, Federal University of Amazonas, Manaus, Brazil

Resent researches during the last three decades showed that the application of natural products have increased with more intensity. A valuable service would therefore be to find materials alternative as fibers to enhanced physic-mechanical properties of cementitious matrices. In this work was evaluated the capability of *Arumã* vegetable fiber with possibilities to reinforce cementitious matrix. The *Arumã (Ischnosiphon polyphyllus),* is found in the Amazon biome. These fibers are widely used by the tradition people of the Northern Region in handicrafts and construction. The objective this work was studied the behavior physic-mechanical of the reinforcement of cementitious matrix from *Arumã* fibers, ciment, metacaulinita and sand. We find that the values of some properties as strength, modulus of elasticity and traction in the cementitious matrix made were compatible with the values of literature and commercial standards. The results showed to highlight potential that exist in this fiber for developing of cementitious matrix

See more of this Session: Non-Thermal Valorization of Waste and Biomass
See more of this Group/Topical: Topical 9: Waste & Biomass Valorization

(21c) Influence of Process Conditions on Properties of Ash From Biomass Gasification

Naomi Klinghoffer, Earth and Environmental Engineering, Columbia University, New York, NY, Marco J. Castaldi, Department of Earth and Environmental Engineering (Henry Krumb School of Mines), Columbia University, New York, NY and Ange Nzihou, Chemical & Environmental Engineering, Ecole des Mines d'Albi-Carmaux, ALBI Cedex 09, France

Influence of process conditions on properties of ash from biomass gasification

Naomi Klinghoffer[1], Marco J. Castaldi[1], Ange Nzihou[2]

[1]Columbia University Department of Earth and Environmental Engineering

[2]Ecole des Mines d'Albi Carmaux, Department of Chemical and Environmental Engineering

Abstract

Biomass is likely to be a significant energy resource in the future. A common way to recover energy from biomass is through gasification where synthesis gas is produced; a by-product of this process is ash. This research investigates the potential to use the ash as a catalyst by understanding the properties of ash generated under different gasification conditions. Specifically, it is desired to produce a porous ash which could be used as a catalyst or as a support for more catalytically active metals.

In this work, poplar wood was gasified under CO_2, steam, and air at different reaction temperatures. Experiments were done in a fluidized bed reactor at temperatures of 750°C and 920°C and ash was recovered. BET-surface area measurements showed that ash from gasification under CO_2 had a higher surface area than ash produced from steam gasification. TGA experiments showed that gasification under CO_2 resulted in a higher mass loss compared to gasification with steam. Gasification with steam/ CO_2 mixtures yielded a mass loss similar to that of steam only which could be indicative of competitive reactions between steam and CO_2.

Introduction

Gasification is a promising technology to yield a usable gaseous product from biomass or waste with high efficiency. There are three main components that are produced from gasification: light gases (such as H_2, CO, and CO_2) that can be used for synthesis of fuels or combustion applications, tars (heavier hydrocarbons), which need to be removed as they can cause problems with downstream equipment, and ash, which is made up of minerals and metals that do not enter into the gas phase. It has been shown that the characteristics of the ash (for example, composition and porosity) can be affected by the reactive gases that are used for gasification(1). The porosity of the ash, combined with high content of metals and minerals, makes ash a good candidate to be used as a catalyst(2). It is common that ash generated from gasification of biomass or waste is either sent to a landfill or used in concrete. Use of ash in catalytic applications would increase its value as well as avoid the need for more expensive catalysts. This research investigates the potential to use the ash as a catalyst by understanding the properties of ash generated under different gasification conditions. Specifically, it is desired to produce a porous ash which could be used as a catalyst or as a support for more catalytically active metals.

Experimental

In this work poplar wood was gasified and analyzed at three different scales. Gasification was done in an in-house built fluidized bed reactor; various flow-through gases were used and results are presented here from experiments done under 10% CO_2 in N_2 and 90% H_2O in N_2. The reactor was heated up to a maximum temperature of 750°C or 920°C and held at the maximum temperature for 1 hour or 30 minutes. The top of the reactor was equipped with a porous frit therefore all ash was trapped inside the reactor and recovered after experiments. BET surface area of the ash samples was measured (Micrometrics Gemini). TGA/DSC (Setaram 111) experiments were done where the biomass samples were heated to 800°C at 20°C/min Experiments were done at different concentrations of CO_2 and H_2O in N_2. Gasification was also done in an environmental scanning electron microscope (FEI XL30) to observe the changes in physical properties of the sample during gasification. Gasification was done under air, CO_2 and H_2O and images were recorded throughout the experiments.

Results and Discussion

After fluidized bed experiments, the ash was recovered and BET-surface area was measured. Results are shown in Figure 1. When comparing surface area of ash generated under steam at 750°C, it is clear that longer gasification time leads to a higher surface area ash. However, mass recovery data shows that the mass of ash recovered is similar in both cases; for 30 minutes 5.60% of the initial mass is recovered as ash and at 1 hour 4.95% of the initial mass is recovered. TGA tests were also done with steam at 800°C (with 9%H_2O in N_2); after 30 minutes at 800°C the mass loss is 80.5% and after 1 hours mass loss is 82.7%. Therefore, both TGA tests at low H_2O concentrations and fluidized bed experiments at high concentrations of H_2O both show that mass loss is very gradual at temperatures in the range of 750-800°C. However, the BET surface area of the ash from fluidized bed experiments showed very different surface areas based on the time spent at maximum temperature. At 1 hour the surface area of the ash sample is 621m^2/g and at 30 minutes the surface area is 429m^2/g. This indicates that it is possible to produce ash with significantly different properties without necessarily influencing overall production of gas/tars. This is important because if it desired to produce ash with specific properties (for example, higher porosity), one would want to do this without impacting overall gas yield.

A.

□□ p>**Figure 1** Ash recovered from fluidized bed experiments: A. surface area B. mass recovery

Gasification was done in the fluidized bed at 750°C for 30 minutes under CO_2 (10% CO_2 in N_2) and H_2O (90% H_2O in N_2) and BET-surface area was measured. The mass recovered was 15.44% under CO_2 and 5.6% under H_2O; it is expected that the mass loss would be higher under steam since more reactant was introduced. However, the surface area of the two samples was the same. This indicates that the CO_2 produces a more porous ash per gram of biomass reacted. Therefore, if it desired to produce a more porous ash, CO_2 is likely a better candidate than H_2O.

TGA tests were done on biomass under 100% CO_2, 9% CO_2 in N_2, 9% H_2O in N_2, and $H_2O/CO_2/N_2$ mix (9%, 8%, balance), as shown in Table 1. The highest mass loss was 98.25% with 100% CO_2. It is reported in the PHYLLIS database that poplar wood has an ash content of 1.5% (dry) therefore the results with pure CO_2 demonstrate almost complete reaction. In the temperature range of 100-200°C, there is dehydration of the sample and a higher mass loss is observed for the dry samples where there is a stronger driving force for water to leave the sample. When 200°C is reached the mass loss under pure CO_2 is 7.11% while the mass loss under 9% CO_2 in N_2 is 5.85%. If dehydration is the only process taking place then one would expect equal mass loss in both cases, since both use dry gases. For example, TGA experiments were done at with air and N_2 at identical conditions (heating rate of 2°C/min to

500°C); both gases were dry. While the final mass loss was very different for the two gases (98.4% for air and 76.6% for N_2), the mass loss was similar until temperatures of 200°C. Therefore, it is possible that under CO_2 reaction begins to take place at temperatures as low as 150°C. ESEM images also showed more physical changes at low temperatures under CO_2 compared to H_2O. For example, at 400°C the pores in the biomass sample have expanded under CO_2 whereas the sample under steam has very similar physical properties to the raw biomass, as shown in Figure 2.

Table 1 Mass loss in TGA gasification experiments

Gas	Final mass loss (%)
100% CO_2	98.25
9% CO_2 in N_2	86.66
9% H_2O in N_2	81.95
9% H_2O, 8% CO_2 in N_2	81.27

Figure 2 ESEM images A. CO_2, raw biomass B. CO_2, 430°C C. H_2O, raw biomass D. H_2O, 425°C

Mass loss was 86.66% under 9% CO_2 in N_2 and 81.95% under 9% H_2O in N_2. The higher mass loss under CO_2 could be partly due to the Boudouard reaction, shown in Equation 1. It was also interesting to note that mass loss under the mixture of CO_2 and H_2O was almost identical than mass loss under H_2O only (81.27% with CO_2 and H_2O and 81.95% with H_2O only). This could be indicative of a competitive reaction between H_2O and CO_2.

Equation 1 $CO_2 + C$ fl $2CO$ *Boudouard Reaction*

Conclusion

If the ash generated from biomass gasification is to be used in catalytic applications it is important to understand the properties of the ash as well as the ash yield that one can expect from different reaction conditions. Overall, it is desirable to produce an ash with catalytic properties without significantly impacting overall gas recovery. Gasification with CO_2 produces a more porous ash compared to gasification under steam. A higher mass loss is observed under CO_2 compared to steam and reactions under CO_2 begin at lower temperatures compared to steam.

References: (1) Butterman, H.C.; Castaldi, M. J. *Environ. Sci. Technol.* **2009**, *43*, 9030-9037

(2) El-Rub, Z. A.; Bramer, E. A.; Brem, G. *Fuel.* **2008**, 87, 2243-2252

See more of this Session: Non-Thermal Valorization of Waste and Biomass
See more of this Group/Topical: Topical 9: Waste & Biomass Valorization

(21d) Bringing Value to Waste Through Gas Fermentation

Michael A. Schultz, LanzaTech Limited, Roselle, IL

Waste gas streams are produced in many process industries, where they are either used as a fuel source, or simply are flared before being released to the atmosphere. Gas streams rich in carbon dioxide, water, carbon monoxide and/or hydrogen also can be produced through combustion or gasification of solid waste streams such as municipal solid waste, forestry residues, or agricultural residues, where the resulting gas stream is used to produce heat and/or power. In both of these cases, the main driver is to reduce the waste stream, or to convert a toxic species such as carbon monoxide to something more benign such as carbon dioxide, as well as to extract value by production of power and/or heat.

Biological fermentation of carbon monoxide rich gas streams provides an alternative route to conversion of these waste streams. The LanzaTech process produces ethanol and other high-value fuel and chemical products from renewable resources that feed on these gas streams. The carbon monoxide rich gas stream can contain any amount of hydrogen, providing an opportunity to process waste streams from a wide range of industries. The brings a significant upgrade in value to the carbon in the waste stream as compared to heat or power production, as well as an opportunity to reuse the carbon rather than release it as CO2 to the atmosphere.

See more of this Session: Non-Thermal Valorization of Waste and Biomass
See more of this Group/Topical: Topical 9: Waste & Biomass Valorization

(21e) Evaluating Different Methods Used to Manufacture Particleboards Produced Using the Blend of Plant Waste From Industrial Processes and Different Resins

Felix Gamarra, Department of Gama, University of Brasilia, UnB Gama College, Brasilia, Brazil, **Valquiria Quirino**, Forestry, Virginia Tech, Blacksburg, VA, Miriam Lima, Department of Technology, Federal University of Amazonas, Brazil, Raimundo Vasconcelos, Department of Technology, Federal University of Amazonas, Manaus and **Edison Bittencourt**, Chemical Engineering, State University of Campinas, Campinas, S„o Paulo, Brazil, Brazil

Solid wastes from industrial processes have the potential of polluting groundwater, the atmosphere, and soils. This makes its final disposal a major environmental concern. A valuable service would therefore be to find alternative uses for this waste. In this work plant waste generated by the industry was used to produce particleboards. Two types of panels were produced. The first used the blend of waste from Cumaru (*Dipteryx odorata*) and polyurethane resin. The second used the blend of waste from wood of Eucalyptus (*Eucalyptus* sp.), asparagus (*Asparagus officinalis*), sugar-cane (*Saccharum officinarum*), and urea-formaldehyde (UF) resin. Physical-mechanical properties of the particleboards, including modulus of rupture, strength, modulus of elasticity, and perpendicular traction, were evaluated following the EN and the ASTM norms. The results obtained in this work were satisfactory. The values found were compatible to those found in the literature and in commercial standards. Thus, this experimental work demonstrated the potential use of renewable resource to produce panel for the civil industry.

Keywords: Vegetal waste, particleboard, physical-mechanical properties, resin.

See more of this Session: Non-Thermal Valorization of Waste and Biomass
See more of this Group/Topical: Topical 9: Waste & Biomass Valorization

(21f) Surface Modification of Various Substrates for Microbial Adsorption Optimization

Lee-At Sabbagh, **Diane Hildebrandt**, David Glasser and T. Matambo, Comps, University of Witwatersrand, Johannesburg, South Africa

Conversion of biomass to useful bio-products is a lengthy and often inefficient process. In our research, we have looked at the conversion of cellulose to ethanol by means of termite bacterial consortium in a fluidized bed bioreactor, where the bacteria are attached as bio-film to an activated carbon or cellulose support. Fairly high conversion rates were achieved in this reactor and the process was fairly robust and flexible. However establishing the bio-film on the activated carbon was a lengthy and fairly chaotic process in terms of the time it took the bacteria to attach to the various substrates (Activated Carbon and Cellulose) and form flocculants. The formation of such flocculants substantially increases the reaction surface and hence should optimize the production of ethanol. Bacteria carry a negative surface charge and hence for increased attraction, the surface charge of the substrate should be modified to be positive. This research, performed initially as batch processes, has shown that with the correct surface charge modifications of the substrate the electrostatic attraction forces between the surface and the bacteria are maximized. As a result of the strong electrostatic attraction forces between the two surfaces the bacteria adsorbs and attaches to the substrate quicker and creates a thicker lining on the surface than seen previously. With these results optimization of flock creation in a fluidized bed bioreactor and its effect on the production of ethanol will be further investigated.

See more of this Session: Non-Thermal Valorization of Waste and Biomass
See more of this Group/Topical: Topical 9: Waste & Biomass Valorization

(27a) Nucleate Boiling Enhancements On Porous Graphite and Micro-Porous and Macro-Finned Copper Surfaces

Mohamed EL-Genk, The University of New Mexico, Albuquerque, NM

Nucleate boiling of dielectric liquids is important in engineering applications such as refrigeration and immersion cooling of high power computer chips and central processor units (CPUs). These liquids are chemically compatible and environmentally friendly and have low saturation temperatures (54 - 64 ^{o}C) at atmospheric pressure. Challenges in cooling high power computer chips using nucleate boiling of dielectric liquids are removing the dissipated thermal power of > 100 W, while keeping the junctions temperature < 85 ^{o}C, and mitigating the effect of the hot-spots developing on the chip surface. Hot spots' local heat flux could be more than 3 times the surface average and induced thermal stresses could reduce service life and increase failure frequency of the chips. Because they are high wetting (static contact angle on most metal and silica surfaces < 5°), undesirable excursion in surface temperature prior to initiating nucleate boiling on smooth copper and silicon surfaces could reach or exceed 25 K.

Numerous investigations have been reported on enhancements of nucleate boiling of dielectric liquids on surfaces with reentrant cavities, micro- and macro-porous surfaces, macro- and micro-structured surfaces, surfaces with micro-porous coating and metal foams. In addition to reducing or eliminating the temperature excursion prior to boiling incipience, the high active sites density for bubbles nucleation increases heat transfer coefficient and Critical Heat Flux (CHF) flux and reduces surface superheat. While CHF is a practical limit in engineering applications, desirable operation is at or below the maximum nucleate boiling heat transfer coefficient; it is much higher than that at CHF and occurs at a lower surface temperature.

Results of experimental investigations on enhancements of saturation and subcooled nucleate boiling of dielectric liquids FC-72, PF-5060 and HFE-7100 on porous graphite and micro-porous and macro-finned copper surfaces are presented, including the effect of surface inclination from 0° (upward-facing) to 180° (downward-facing). Results are compared with those reported by other investigators for macro-structured and micro-porous surfaces and surfaces with micro-porous coatings. In addition, results of CFD calculations are presented of sizing copper spreaders with thin surface layers of porous graphite and micro-porous copper for cooling 10 x 10 mm underlying chips using nucleate boiling of FC-72 and PF-5060 liquids.

See more of this Session: Donald Q. Kern Memorial Award Lecture
See more of this Group/Topical: Energy and Transport Processes

(35a) Power System Design for a Hybrid Fuel Cell Vehicle: A Globally Optimal Approach

Syed Kaschif Ahmed and Donald J. Chmielewski, Chemical and Biological Engineering, Illinois Institute of Technology, Chicago, IL

The notion of a fuel cell powered vehicle has captured the imagination of many as a clean/efficient alternative to the internal combustion engine. However, the capital cost of a fuel cell power unit is significant, and threatens to be a show-stopping hurtle. As such, many have looked to hybridize the fuel cell with lower cost energy storage devices; a rechargeable battery and / or super-capacitor. These low energy / high power density devices will provide supplemental power during periods of large demand and thus allow for the selection of a smaller / lower cost fuel cell unit. The basic design challenge is to size of the fuel cell and supplemental energy storage units so as to minimize overall cost.

Due to the drive cycle expected during vehicle operation, the proposed design scheme should include power demand variability as an underlying element. It is further recognized that a high level controller is needed to coordinate device interactions as well as ensure that demanded power is sent to the electric motor driving the vehicle. Clearly, the characteristics of this high level controller will influence the required size of each device. As such, controller characteristics have been included as design variables. Additional design criteria include; max power and energy density for each type of device, losses associated with energy conversion, friction losses, the impact of device size on vehicle mass and overall power demand.

Concerning computational issues, we will show that nearly all of the nonlinear constraints discussed above can be converted to a convex form. The exception is a limited set of scalar reverse-convex inequalities, used to define the relation between variance and standard deviation for key process variables. Given this convex / reverse-convex form a computationally efficient branch-and-bound algorithm has been developed to obtain globally optimal solutions to the proposed hybrid vehicle design problem.

See more of this Session: Recent Advances in Fuel Cell and Battery Technologies
See more of this Group/Topical: Energy and Transport Processes

(35b) Testing of Electrode Durability for Extended Operation within An Alkaline Fuel Cell

Matthew S. Naughton, Department of Chemical & Biomolecular Engineering, University of Illinois at Urbana-Champaign, Urbana, IL and Paul J. A. Kenis, Department of Chemical and Biomolecular Engineering, University of Illinois at Urbana-Champaign, Urbana, IL

Fuel cells hold promise as alternative power sources due to their ability to bypass Carnot efficiency limitations by directly converting chemical energy into electrical energy. However, the high costs of catalysts and membranes, as well as component durability issues, have barred widespread implementation. Alkaline fuel cells (AFCs) have gained increasing attention due to superior electrode kinetics in alkaline media, compared to acidic media, which enables the use of non-noble metal catalysts [1]. However, alkaline fuel cells are susceptible to flooding at the anode, which blocks gas diffusion and reduces catalyst utilization. PTFE-bonded electrodes are standard in alkaline media, for both hydrophobicity to reduce flooding and stability; however, the sintering process required to produce these electrodes increases catalyst particle size, resulting in a decrease in catalytic activity [1].

Previously, we have developed a pH-flexible flowing electrolyte microfluidic fuel cell, which uses an external reference electrode to individually analyze cathode and anode performance. This microfluidic configuration combines the versatility of a traditional three electrode cell with the conditions found in an operating fuel cell. To date, typical analyses have consisted of short term experiments with a syringe pump supplying a liquid electrolyte. Although this setup is valuable for short term experiments, long term experiments require a recirculating electrolyte, which is susceptible to dilution through water formation. In addition, water buildup on the anode can cumulatively hinder the performance of the cell due to electrode flooding. The effect of extended operation is not easily simulated using syringe pump experiments alone.

Here, we present long-term experimentation exploring the use of a hydrophobic additive in alkaline media. The fuel cell is operated with a recirculating electrolyte, and full polarization curves are taken at regular intervals over the course of hours, revealing a full spectrum of insight into cell performance over time. The use of the external reference electrode allows us to determine individual electrode performance over time, and the extent of degradation is quantified.

[1] Brushett et al., Journal of the Electrochemical Society, 2009, 156, B565

See more of this Session: Recent Advances in Fuel Cell and Battery Technologies
See more of this Group/Topical: Energy and Transport Processes

(35c) A Computational Simulation of the Performance of a Zinc-Air Fuel Cell with Operational Condition

Jungyun Kim Sr., Sangmin Park Sr., Taeyoung Oh and Hoil Lee, Energy Department, R&D Division, Hyundai Industrial Research Institute, Ulsan, South Korea

The numerical analysis for the electrochemical performance with operational conditions was carried out for a zinc-air fuel cell(ZAFC). The Nernst-Planck equation with a boundary condition of Butler-Volmer type was adopted to describe electrochemical effects of mass transfer, migration, kinetics of electrode. The Navier-Stokes equation, coupling to the Nernst-Planck equation, is also applied to describe external air flow fields. Polarization curves had been investigated under various operational conditions, such as air flow field, electrolyte concentration and temperature. It has been derived several design factors fo a ZAFC from our study to maximize the cell performance.

See more of this Session: Recent Advances in Fuel Cell and Battery Technologies
See more of this Group/Topical: Energy and Transport Processes

(35d) Simulations of Air-Independent Solid Oxide Fuel Cell System for Underwater Applications

Taeyoung Oh, Energy Research Department, Hyundai Industrial Research Institute, R&D Division, Hyundai Heavy Industries, Co., Ltd., Ulsan, South Korea and Hoil Lee, Energy Research Department, R&D Division, Hyundai Heavy Industries, Co., Ltd., Ulsan, South Korea

To investigate the possibility of solid oxide fuel cell system for the underwater applications, the steady and dynamic analysis of the system were simulated. The fuel cell stack, methanol steam reformer, fuel and oxidant storage, and other balance-of-plants were integrated. The selected reactants were methanol as fuel and liquid oxygen as oxidant. The operation conditions of the fuel cell stack were 480kW power output, the temperature of 950°C, and the pressure of 2 bar. The dynamic behaviors of the oxygen tank were also examined, including the tank pressure, temperature, and the required heat to vaporize the liquid oxygen. This process simulation results led to derive the required characteristics and performances of the main components for the underwater applications.

See more of this Session: Recent Advances in Fuel Cell and Battery Technologies
See more of this Group/Topical: Energy and Transport Processes

(35e) A Novel Synthesis Method for Carbon-Supported Platinum Nanoparticles for Proton Exchange Membrane Fuel Cell

Jong-Sung Yu, Min-Sik Kim, Jung Ho Kim, Min Young Song and Baizeng Fang, Advanced Materials Chemistry, Korea University, Jochiwon, South Korea

A simple and efficient approach has been developed for synthesis of carbon-supported Pt nanoparticles (NPs) that combines homogeneous deposition (HD) of Pt complex species through a gradual increase of pH realized by *in situ* hydrolysis of urea and subsequent uniform reduction by ethylene glycol (EG) in a polyol process, giving control over the size and dispersion of Pt NPs. With increasing amount of urea in the starting Pt salt aqueous solution, the size of Pt complex species decreases and so does that of the metallic Pt NPs. The excellent electrocatalysis ability of the Pt catalysts produced by HD-EG is demonstrated through the determination of electrochemical surface area and fuel-cell polarization performance. The Pt NPs deposited on Vulcan XC-72 (VC) carbon black by the HD-EG strategy show smaller size with more uniform dispersion, higher Pt utilization efficiency, and considerably improved fuel-cell polarization performance compared with the Pt NPs prepared by conventional sodium borohydride reduction or by a microwave-assisted polyol approach. Particularly important and significant is that this HD-EG method is very efficient for the synthesis of high Pt loading catalysts with tunable NP size and uniform particle dispersion. A high metal loading catalyst such as Pt(60 wt %)/VC fabricated by the HD-EG method outperforms ones with mid-to-low metal loadings (i.e., 40 and 20 wt %), even at a very low catalyst loading of 0.2 mg of Pt cm^{-2} at the cathode, which is for the first time reported for the VC-supported Pt catalysts.

See more of this Session: Recent Advances in Fuel Cell and Battery Technologies
See more of this Group/Topical: Energy and Transport Processes

PVA-g-PAA Substituted Benzimidazolium Group Membrane for Fuel Cell Applications

Pratima Gajbhiye, Pranjal Joshi, Jayant K.Singh and Anil Kumar, Department of Chemical Engineering, Indian Institute of Technology Kanpur, India

1. Introduction:

Fuel cells have typically been classified according to their working temperature. The six common types of Fuel cells are alkaline fuel cell (AFC), proton exchange membrane fuel cell (PEMFC), phosphoric acid fuel cell (PAFC), direct methanol fuel cell (DMFC), molten carbonate fuel cell (MCFC), and solid oxide fuel cell (SOFC). Historically, the most widely used membrane has been Nafion 117 from DuPont and considerable amount of literature is dedicated on Fuel cells with this. In an effort to find the material other than this, researchers have focused on the thermally stable and conducting polymers and in this connection, they have concentrated upon the three classes of membranes: Polybenzimidazoles (PBI), Polyimides and Sulfonated polymer membranes.

Out of various membranes reported in the literature, we have chosen to synthesize grafted PVA type membranes for our studies. The grafting of this material has been carried out using Acrylic acid, maleic anhydride and methacrylic acid. The membranes thus formed have been used to covalently bind an ionic liquid involving benzimidazolium groups. The synthesis of various monomers needed for this modification is given in Table 1. The grafted membranes were crosslinked with Compound I and Compound II to form benzimidazole substitution. The membranes thus prepared were characterized by standard techniques such as MWCO, SEM, AFM, TGA, IEC, and FT-IR, Contact angle measurement, ion-exchange capacity, water uptake capacity and rejection of chromic acid. These crosslinked polymers were then used in the Fuel cell and results were compared against Nafion-117.

The membranes were tested in laboratory made Fuel cell set up. The Fuel cell bipolar plates were made of SS316, which was fabricated using Computed numerically controlled machines to get serpentine channels on anode and parallel channels on cathode side. We have synthesized Pt/Ru (anode) and Pt (cathode) catalysts loaded on carbon by immobilization technique. We have also determined the performance of the Fuel cell using Nafion 117 membrane under the same parameters. Experimental results show that the membranes synthesized by us are superior to Nafion 117.

2. Experimental:

2.1 Synthesis of Bis-3-Amino-4-[3-(triethylammoniumsulfonato) phenylamino]-phenyl sulfone (BASPAPS) (reaction 3 of Table 1)

Bis-4-chlorophenyl sulfone (25 g, 85 mmol) and 125 ml of 96 % sulfuric acid were added to a one-liter conical flask (equipped with a reflux condenser) and stirred at room temperature until there was complete solubilization of bis-4-chlorophenyl sulfone. The mixture was then cooled in an ice water bath. A sulfonitric solution was prepared with 70 % fuming nitric acid (16 ml, 220 mmol) and 25 ml of 96 % sulfuric acid, which was added drop wise into the flask. Upon completion of addition, the solution temperature was slowly increased to room temperature and was left to stand overnight. The final heterogeneous reaction mixture containing pale yellow precipitate was slowly poured in 2 litre of de-ionized water. The yellow precipitate was filtered, washed twice with water, and with 200 ml of 1M potassium carbonate solution to eliminate the residual acid. Then the crude product was washed again with water until neutral pH and finally dried for 5 hours under vacuum at 80°C, which yielded 29.4 g of bis-4-chloro-3-nitro-phenyl sulfone.

Further, bis-4-chloro-3-nitro-phenyl sulfone (25 g, 66 mmol), metanilic acid (28.7 g, 165 mmol), triethylamine (45 ml, 230 mmol) and DMSO (120 ml) were added in a 500 ml conical flask (equipped with a reflux condenser). The red solution obtained was heated to 120°C for 6 hours and then precipitated in 500 ml of ethyl acetate. The yellow powder obtained was filtered from the solution and recrystallized in ethanol which yielded 21.4 g of bis-3-nitro-4-[3-(triethylammoniumsulfonato) phenylamino]-phenyl sulfone. 50% hydrazine hydrate solution (200 ml) was added to bis-3-nitro-4-[3-(triethylammoniumsulfonato) phenylamino]-phenyl sulfone which was then allowed to stand for 2-3 hours. Then a 100 ml of dilute hydrochloric acid (1:1) solution was added to it and the solution was filtered giving 13 g of bis-3-amino-4-[3-(triethylammoniumsulfonato) phenylamino]-phenyl sulfone (BASPAPS) on drying.

2.2 Synthesis of PVA-g-PAA/BASPAPS membrane

10 g of polyvinyl alcohol (having molecular weight of 1, 30, 000 Dalton) was dissolved in water (90 ml) by stirring at 90°C for 6 hours. Then acrylic acid

(7.2 g, 0.1 mol) and 1.5 g of Bis-3-Amino-4-[3-(triethylammoniumsulfonato) phenylamino]-phenyl sulfone (BASPAPS) dissolved in 10 ml of water was added to the mixture and the reaction was allowed to proceed at 80°C for 5 hours which resulted in a homogeneous solution. Now the solution obtained is highly hydrophilic due to hydroxyl groups of PVA and they impart tendency of the membrane to swell in water. This was suppressed by adding a specific amount of Glutaraldehyde to the solution and then it is cast on a glass plate at room temperature for drying. The membrane was further immersed in 5% glutaraldehyde solution with 1% HCl solution and acetone to further crosslink the membrane. This was then heated in an oven for one hour at 180°C. The obtained membrane was immersed in 20%Orhtophosphoric acid to make it acidic in nature. Then it was washed with deionized water and further dried in atmospheric air to make Membrane electrode assembly (MEA) which is the heart of the Fuel cell.

2.3 Preparation of MEA

The electrodes were made of carbon cloths that were cut into the required size and the catalyst was applied with the polymer ink synthesized in laboratory along with isopropyl alcohol and Ethanol of required proportion using ultrasonic water bath to produce catalyst slurry. It was then spread on the cloth as wet film using paint-brush technique. They were then allowed to dry at 80°C for 1hour so that a proper binding of the catalyst on the electrode occurs. The catalyst loading was 2mg/cm^2 for Pt-Ru and 4mg/cm^2 for Pt and the MEA was made by pressing at

120kg/cm^2 at temperature of 180°C for 3min. The Toray Carbon Paper, TGP-60 was used as the gas diffusion layer on each side of the electrode.

3. Results and discussion

FTIR, NMR and CHN analysis were used to characterize compounds synthesized in this work and they confirm their molecular structures given in Table 1.Various membranes were analyzed by MWCO experiments, and 90% rejection of the solute occurred at 23500 and 25000 molecular weight. These correspond to the average pore size of 9.10nm and 9.42nm respectively. The SEM images of the top surfaces of the membrane show white spots in black background on the surface which seems to imply a second phase on the surface. The TGA analysis shows about 5 and 10 % weight loss upto 170 and 200°C and further weight loss is due to degradation of BASPAPS. This reaction occurs at 480°C leading to total disintegration of the membrane. The 5-10% loss is due to loss of both free and bound water molecules and is consistent with the results reported in literature. In order to estimate the wall charge of the ultrafiltration membrane, we have carried out the separation of chromium (VI) solution. The flow through the capillaries was assumed to follow the Space charge model and the analysis and the curve fitting of the data showed the wall potential to be around 525mV. The analysis procedure has been reported in our earlier works and shows that the polymers reported in this work are highly charged. Subsequently we prepared the MEA and using methanol as a fuel, we experimentally evaluated the function of the fuel cell. We also determined the performance of Nafion-117 membrane and compared our results in Figure 1. The superiority of our membrane is clearly seen and the region of large current is found to be almost horizontal indicating very low internal resistance of the Fuel cell. This behavior confirms the superiority of our membrane.

4. Conclusion

The studies on PVA grafted membrane reported in the literature have shown that the methanol leakage is minimal and the starting point of this work has been this class of material for membrane synthesis. A new class of ionic liquid has been synthesized by modifying our membrane with benzimidazolium group. Fuel cell studies show that this membrane gives superior results compared to Nafion-117.

Table 1. Synthesis of various monomers required for PBI polymers

Table 1:

(1) Synthesis of Diaminobenzidine hydrochloride (DAB) from Benzidine [1]:

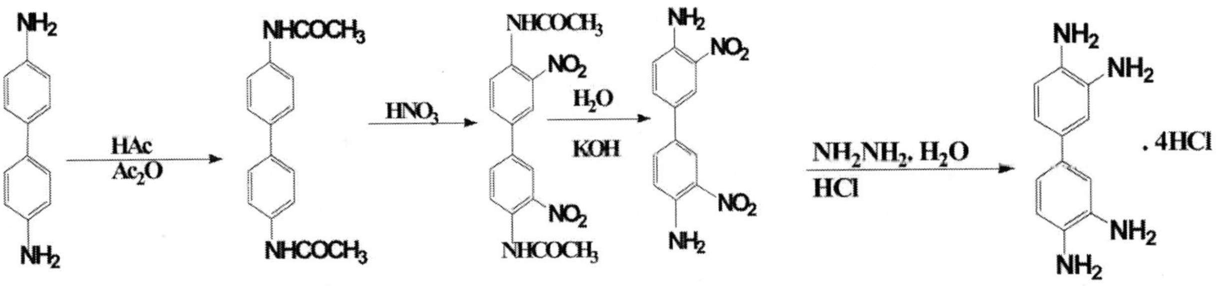

(Benzidine) (DAB)

(2) Synthesis of Diaminobenzidine sulfone hydrochloride (DABS) from Diaminodiphenyl sulfone (DADS):

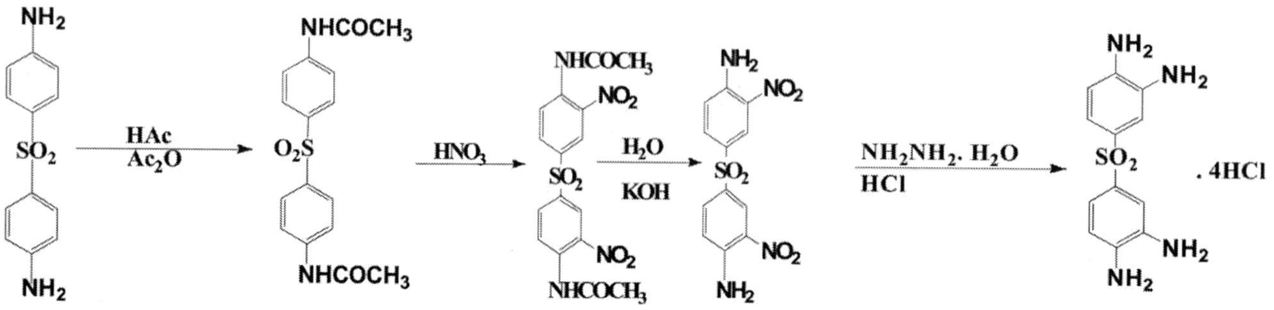

(DABS) (DADS)

(3) Synthesis of Bis-3-Amino-4-[3-triethylammoniumsulfonato)phenylamino]-phenyl sulfone (BASPAPS) from Dichlorodiphenyl sulfone (DCDPS)[2,3]:

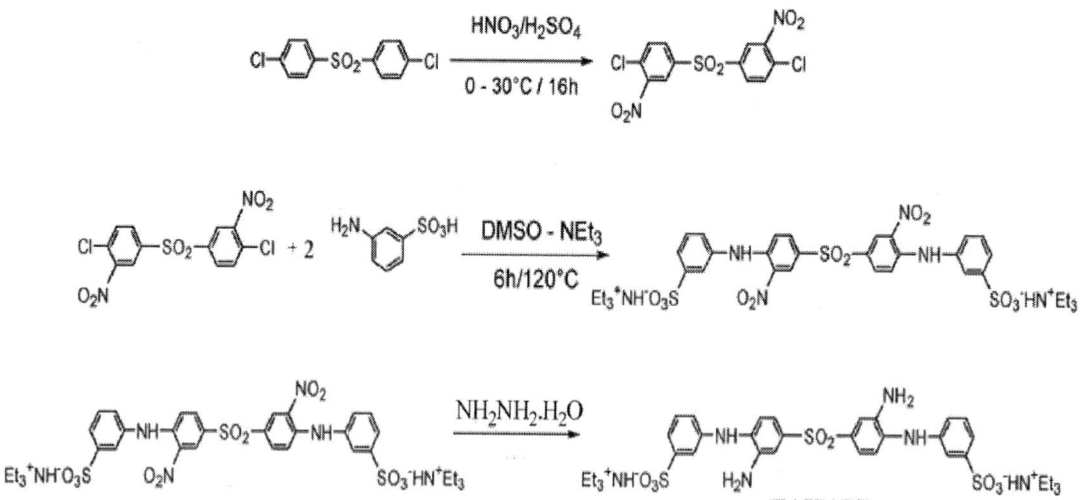

(BASPAPS)

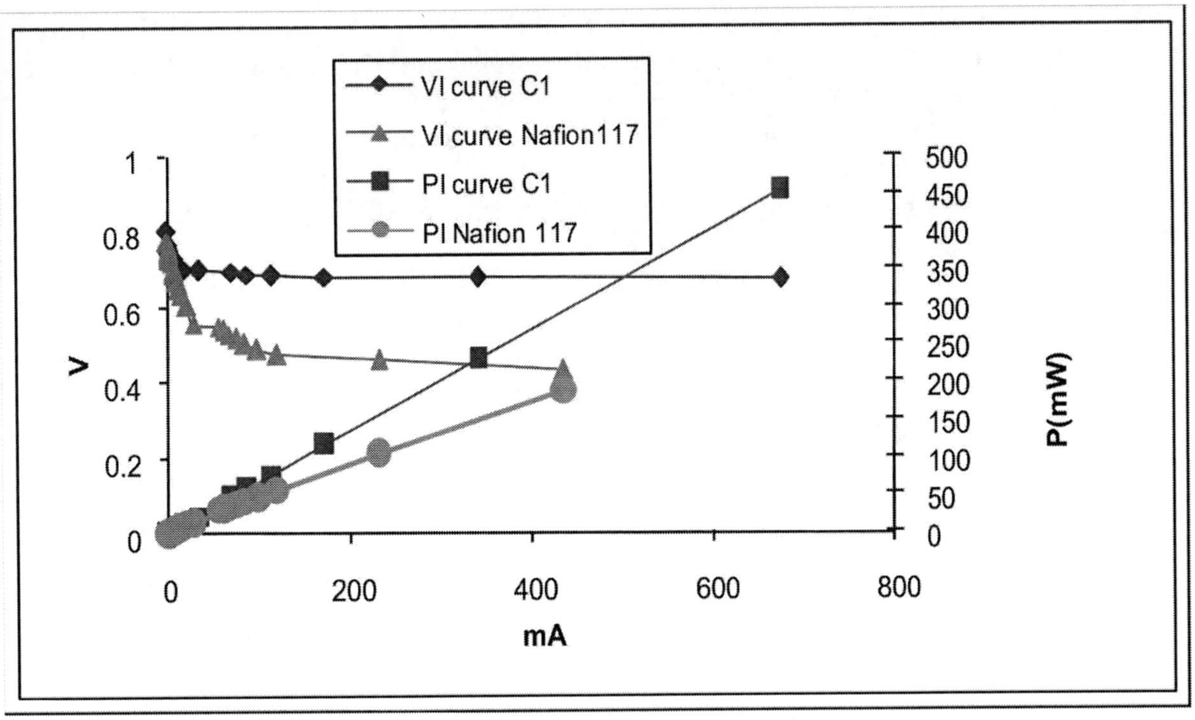

Figure 1. V-I and Power characteristics of the Fuel cell used in our studies. membranes reported in this work gives higher performance.

1) J. Hoste, Diaminobenzidine as a reagent for vanadium and selenium, Analytica Chimica Acta 2 (1948) 402

2) J. Jouanneau, R. Mercier, L. Gonon, and G. Gebel, Synthesis of Sulfonated Polybenzimidazoles from Functionalized Monomers: Preparation of Ionic Conducting Membranes, Macromolecules 40 (2007) 983

3) J. Jouanneau PhD Thesis, Préparée au Laboratoire des Matériaux Organiques à Propriétés Spécifiques (LMOPS) et dans le groupe Polymères Conducteurs Ioniques (SPrAM), 2006 (source: http://tel.archives-ouvertes.fr)

The relation between ink agglomerate size and pore size distribution of the cathode catalyst layer of PEM fuel cells and the effect of carbon corrosion on CCL structure

M. Baghalha[a], and M. Eikerling[b]

[a] Department of Chemical and Petroleum Engineering, Sharif University of Technology, Tehran, Iran

[b] Department of Chemistry, Simon Fraser University, Burnaby, BC, V5A 1S6, Canada

Carbon corrosion is a major degradation mechanism in cathode catalyst layers (CCL) of polymer electrolyte membrane (PEM) fuel cells. Carbon corrosion (coupled with ionomer dissolution/degradation) induces severe changes in the CCL structure. In the present study, the inter-relation between the particle size of a packed sphere medium and the pore size distribution (PSD) is established. This relation is then used to predict the agglomerate size of the mix of Pt/C and ionomer in the catalyst ink, using experimentally measured PSD of the CCL. During cell degradation, the size of these agglomerates decreases; since, carbon corrosion causes carbon particles to decrease in size. Furthermore, as carbon corrosion proceeds, the ionomer to carbon ratio of agglomerates increase. Literature data suggests that as the ionomer to carbon ratio of the ink increases, the pore sizes and the porosity of the CCL decrease. These data are implemented in the present work to predict the PSD of the CCL at any given state of carbon corrosion.

Introduction

Carbon in the cathode catalyst layer (CCL) of a polymer electrolyte fuel cell (PEFC) undergoes electrochemical oxidation to the corrosion product, carbon dioxide (reaction (I)), with oxygen reduction reaction (reaction (II)) as the coupling reduction reaction.

$$C + 2H_2O \square \; CO_2 + 4H^+ + 4e^- \qquad \text{(I)}$$

$$4H^+ + 4e^- + O_2 \rightarrow 2H_2O \qquad \text{(II)}$$

Carbon corrosion of the CCL can occur very rapidly at high cathode potentials (~1 to 1.5 V). Such high potentials may be generated during PEFC start-up and shut-down [1]. The catalytic effect of Pt could render carbon corrosion rates appreciable even at lower potentials in the operating range (~0.6 to 0.9 V) of PEFC [2,3]. High rates of carbon corrosion can cause severe performance losses of PEFC [4]. This severe performance loss could involve a variety of properties that change with carbon corrosion, including the change in the CCL structural parameters, such as PSD. It was the objective of this work to predict the PSD of the CL at any given state of carbon corrosion.

In this article, the PSD of a packed sphere medium is related to the sphere particle size. A mathematical function for these PSDs is proposed. The PSD of one of the literature works is then investigated under different preparation conditions. From this study, the average agglomerate sizes of the ink at different ionomer to carbon ratios are estimated. Finally, the effect of carbon corrosion on the agglomerate size and structure is discussed.

PSD of a packed medium of uniform size hard spheres

The pore size distributions of packed hard sphere media at three different sphere radius sizes (160, 206, and 410 nm) have been measured by Smith et al., [5]. Using the corresponding sphere radius size for each data set, these PSDs were made dimensionless. The assumed sphere radius size for the calculations were slightly different (~10%) than the values reported by Smith et al [5]. It is believed that this is due to experimental error. The dimensionless PSDs were reasonably close to each other; hence, a dimensionless distribution function could be used to represent this PSD. The differential pore size distribution of each peak was assumed to be represented by the log normal distribution function in Eq. (1),

$$\frac{dV}{dz} = \frac{\varepsilon}{z\sigma\sqrt{2\pi}} \exp[-\frac{1}{2\sigma^2}(\ln\frac{z}{z_M})^2] \quad (1)$$

Where ε, σ, and z_M are the porosity, standard deviation, and the mean pore size, respectively. z is the dimensionless pore diameter, defined as

$$z = \frac{r}{R_{particle}} \quad (2)$$

Where r and $R_{particle}$ are the dimensional pore radius and the particle radius of the packed sphere bed, respectively. The values of the mean and the standard deviation of this function were varied for each peak to obtain a close match, as shown in Figure 1.

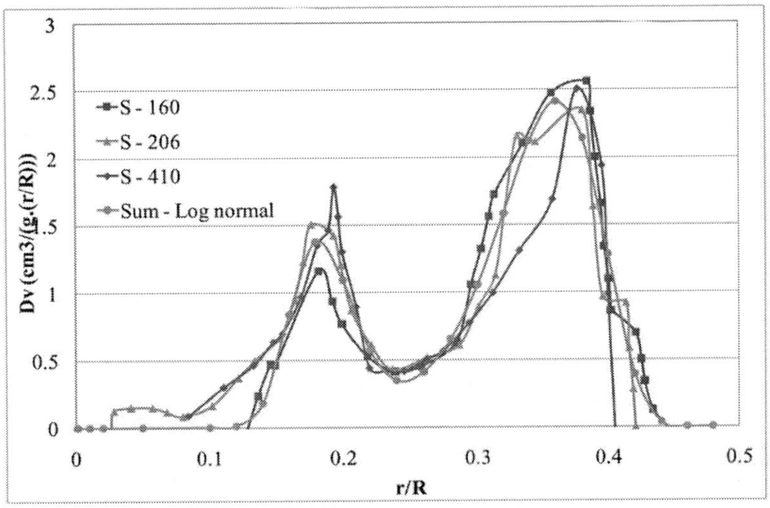

Figure 1. Two log normal distribution functions were used to model the PSDs of the hard spheres media.

The equations for the small (Eq. (3)) and the large (Eq. (4)) peaks are

$$\frac{dV}{dz} = \frac{0.08}{z(0.129)\sqrt{2\pi}} \exp[-\frac{1}{2(0.129)^2}(\ln\frac{z}{0.185})^2] \quad (3)$$

$$\frac{dV}{dz} = \frac{0.25}{(0.5-z)(0.287)\sqrt{2\pi}} \exp[-\frac{1}{2(0.287)^2}(\ln\frac{(0.5-z)}{0.15})^2] \quad (4)$$

The effect of Ionomer to Carbon ratio on the PSD of CL

The PSD data, as measured through MIP by Yu and Carter [6] were digitized using a digitizing software. The data were then normalized, knowing the porosity for each CL with the given I/C ratio. As shown in Figure 2, by increasing I/C ratio, the PSD shifts to smaller sizes and the porosity also decreases (the area under each curve). It is believed that the source of the effect of I/C ratio is in the ink making process. It is envisioned that by increasing I/C ratio that acts as a surfactant also, the size of Pt/C/ionomer agglomerate in the ink decrease. The effect of surfactants on reducing the agglomerate size is well established [7]. Through analogy with the PSD of hard spheres, it is obvious that when the agglomerates are smaller, the pores between them becomes smaller too, as shown in Figure 2.

Through our proposed inter-relation between the PSDs of hard spheres and CL agglomerates, the agglomerate size for each I/C data set could be predicted in such a way that their peaks fall on the major peak of the hard sphere peak. The calculated agglomerate diameter sizes for each I/C data set are reported in Table 1. The dimensionless PSDs for these 3 data sets are shown in Figure 3; as compared to the dimensionless log normal PSD of packed spheres. As shown in this Figure, the overall shapes of these peaks are the same. Only the areas under these curves are different which is due to different values of porosities (see Table 2) at different values of I/C ratio.

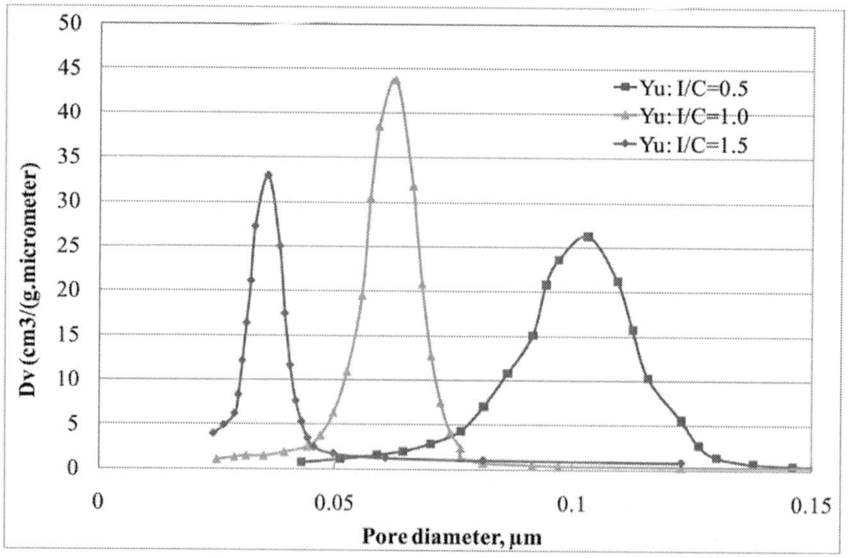

Figure 2. Pore size distribution of the catalyst layer at 3 different I/C ratio.

Table 1: The agglomerate diameter size for 3 I/C ratios calculated based on the PSD data of Yu and Carter [6]

I/C ratio	Calculated agglomerate diameter, nm
I/C=0.5	285
I/C=1.0	175
I/C=1.5	100

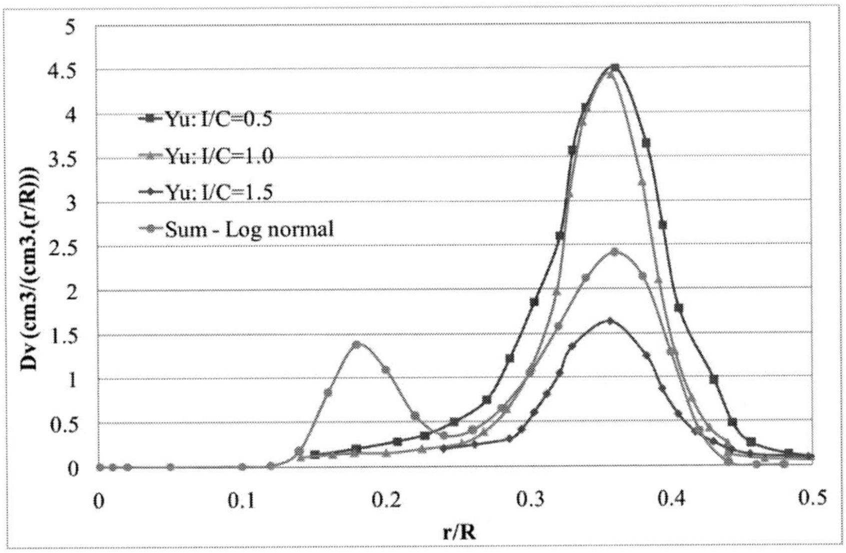

Figure 3. The dimensionless PSDs for the 3 I/C data sets, as compared to the dimensionless log normal PSD of packed spheres.

A correlation between the agglomerate size and PSD as well as the porosity of the CL

There is one major assumption. The basis for this assumption is the following observation by Yu and carter [6]: "Given that the electrode thickness is observed to be independent of the I/C ratio bt SEM, increasing the I/C ratio from 0.5 to 1.5 triples the ionomer volume fraction and correspondingly reduces the electrode void fraction". The following table from the same source [6] was also a key factor in our calculations.

Table 2: Measured and calculated porosities as reported by Yu and Carter [6]

I/C ratio	Measured porosity by MIP method for pore diameters>30 nm (7)	Calculated porosity from densities
Carbon Powder	0.83	N/A
I/C=0.5	0.50	0.73
I/C=1.0	0.42	0.65
I/C=1.5	0.20	0.57

Based on the calculated porosities reported in Table 2, it was found that a "loose agglomerate" model (it means not a tight arrangement between agglomerates) with an inter-agglomerate and an intra-agglomerate porosity of 0.58 fit all the porosity data very well.

The only other parameter in Eq. (1) that needs quantification is ε, i.e., the porosity of un-trapped inter-agglomerate pores. As shown in Figure 4, it was found that a power law function of trapped inter-agglomerate porosity vs. the agglomerate size fits the data well.

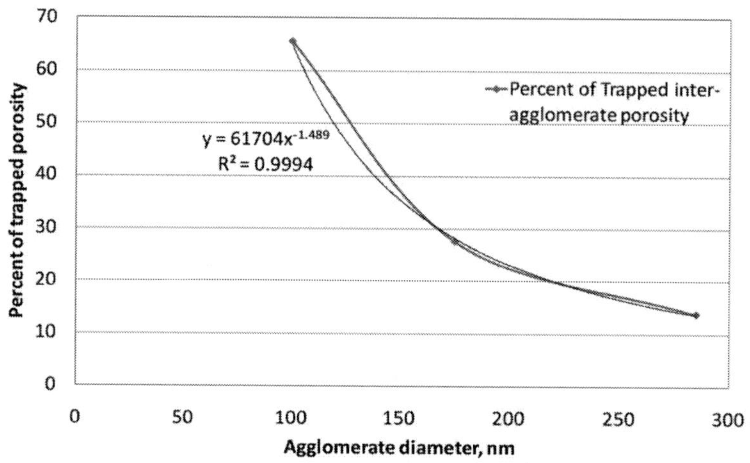

Figure 4. Percent of Trapped inter-agglomerate porosity.

Considering that the measured inter-agglomerate porosity of the catalyst layer can be represented by the larger peak of the log normal function in Figure 1, Eq. (4) is modified as follows to reflect the fact that some portion of the porosity is trapped.

$$\frac{dV}{dz} = \frac{(0.58-y)}{(0.5-z)(0.287)\sqrt{2\pi}}\exp[-\frac{1}{2(0.287)^2}(\ln\frac{(0.5-z)}{0.15})^2] \tag{5}$$

Where y and z are

$$y = 617(2R_{agglomerate})^{-1.489} \tag{6}$$

$$z = \frac{r}{R_{agglomerate}} \tag{7}$$

As a result, the only parameter that affects PSD of the larger inter-agglomerate pores in Eqs. (5) to (7) is the agglomerate radius, $R_{agglomerate}$. These calculations were based on the assumption that the major cause of obtaining different PSDs for different I/C ratio in Figure 2 was the effect of additional ionomer content on reducing the agglomerate diameter.

Carbon corrosion and its effect on the change in agglomerate size and PSD during degradation

As a case study in this section, it is assumed that a CCL with an I/C=1.0 is in normal operation with carbon corrosion occurring under mild condition; i.e., low cathode potential that normally produces low carbon corrosion rate that do not severely reduce the CCL thickness over time. Under such normal conditions, it is assumed that carbon corrosion does not damage the "loose" structure of the agglomerates, i.e., the porosity remains constant. As a result, upon corrosion and reduction of the carbon volume inside an agglomerate, the new agglomerate size at time t, $d^t_{\text{agglomerate}}$, can be obtained according to

$$d^t_{\text{agglomerate}} = d^i_{\text{agglomerate}} \left(0.42[1.381\ \text{CFR} + 1]\right)^{1/3} \tag{8}$$

Where **CFR** is the carbon fraction remained. Eq. (8) was derived for the case of I/C=1.0 and assuming that the volume ratio of the porosity phase to the carbon phase remains constant at any given time during carbon corrosion; i.e.,

$$\frac{X^t_p}{X^t_c} = \frac{X^i_p}{X^i_c} = \frac{0.16}{0.42} = 0.381 \quad \text{(when I/C=1.0)} \tag{9}$$

Where X^t_p and X^t_c are the volume fractions of pores and carbon, respectively. Assuming mild carbon corrosion, the value of CFR was changed from 1 to 0.6. Obviously, the change in the agglomerate diameter is not significant. Using the values of the agglomerate diameter at 3 CFR values, the PSD of the inter-agglomerate porosity can be calculated from Eqs. (5) to (7). The evolution of these PSDs as carbon corrosion proceeds up to 40% of the initial carbon volume is shown in Figure 5.

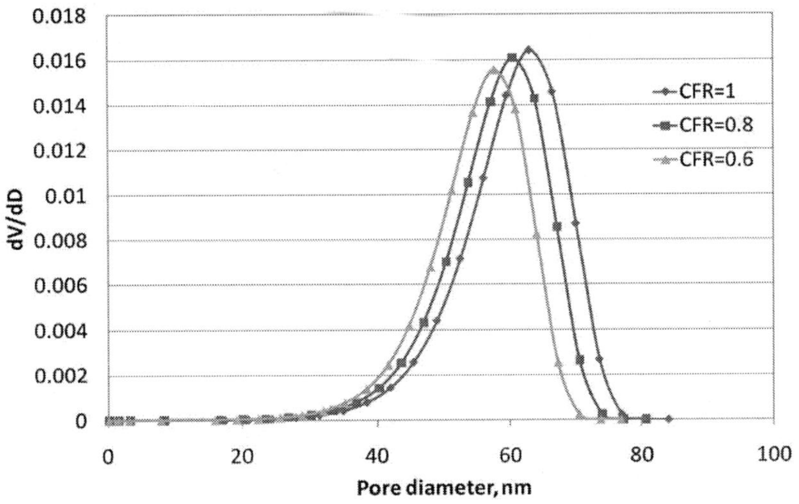

Figure 5. The evolution of the PSDs of the inter-agglomerate porosity as carbon orrosion proceeds up to 40% of the initial carbon volume. This is a case study for I/C=1.0, when $d^i_{\text{agglomerate}} = 175\ \text{nm}$.

Conclusions

The inter-relation between the particle size of a packed sphere medium and the pore size distribution (PSD) was established using a dimensionless log normal function. This relation was applied to predict the agglomerate size in the catalyst ink in one case study, using experimentally measured PSD of the CCL. During cell degradation, the size of these agglomerates decreases; since, carbon corrosion causes carbon particles to decrease in size. Furthermore, as carbon corrosion proceeds, the ionomer to carbon ratio of agglomerates increase. Literature data suggests that as the ionomer to carbon ratio of the ink increases, the pore sizes and the apparent (un-trapped) porosity of the CCL decrease. These data were implemented in the present work to predict the PSD of the CCL at any given state of carbon corrosion.

References

1. J. Kim, J. Lee, Y. Tak, *Journal of Power Sources*, **192**, 674 (2009).
2. W. Li, A.M. Lane, *Electrochemistry Communications*, **11**, 1187 (2009).
3. V.A.T. Dam, K. Jayasayee, F.A. deBruijn, *FUEL CELLS*, **09**, 453 (2009).
4. A.P. Young, J. Stumper, E. Gyenge, *Journal of The Electrochemical Society,* **156**, B913 (2009).
5. D.M. Smith, et al, Powder Technology, 53, 11-22 (1987).
6. Z. Yu, R. N. Carter, Measurements of Effective Oxygen Diffusivity, Pore Size Distribution, and Porosity in PEM Fuel Cell Electrodes, *ECS Transactions*, **19** (17), 1 (2009).
7. C.-N. Chen, Y.-L. Chen, W.J. Tseng, Surfactant-assisted de-agglomeration of graphite nanoparticles by wet ball mixing, *Journal of Materials Processing Technology*, **190**, 61-64 (2007).

(40a) Energy and Greenhouse Gas Emissions of Pyrolysis Pathways

Jeongwoo Han, Amgad Elgowainy, Ignasi Palou Rivera, Michael Wang and Jennifer M. Dunn, Energy Systems Division, Argonne National Laboratory, Argonne, IL

Pyrolysis of biomass to produce liquid fuels can play a significant role toward achieving the goal of 36 billion gallons per year of biofuels by 2022 as set by the Energy Independence and Security Act (EISA) of 2007. Argonne National Laboratory conducted life cycle analysis of various pyrolysis pathways by expanding and employing the Greenhouse Gases, Regulated Emissions, and Energy Use in Transportation Model (GREET). The fuel cycle energy use and greenhouse gases (GHG) emissions from the pyrolysis pathways were compared to the baseline gasoline pathway as well as alternative biofuel pathways such as corn and cellulosic ethanol, cellulosic biomass to Fischer Tropsch Diesel (FTD), and soybean biodiesel, renewable diesel and renewable gasoline. The life cycle results are significantly impacted by the source of hydrogen production for hydrotreatment and hydrocracking processes, and the handling of the biochar and other coproducts of the pyrolysis pathway.

See more of this Session: Advances In Alternative BioEnergy
See more of this Group/Topical: Topical 9: Waste & Biomass Valorization

(40d) Efficient Fermentation of Glucose and Xylose Mixtures Using Immobilized Saccharomyces Cerevisiae on Pine Wood Chips and Free Pichia Stipitis

Ravi Dhabhai, Chemical Engineering, Malaviya National Institute of Technology Jaipur, Jaipur, India, **S. P. Chaurasia**, Chemical Engineering, Malaviya National Institute of Technology, Jaipur, India and Ajay K. Dalai, Chemical Engineering, University of Saskatchewan, Saskatoon, SK, Canada

In this work, the ability of pine wood chips (PWC) as yeast support has been evaluated. *Saccharomyces cerevisiae* was immobilized on this biological support and mixed sugar fermentation was carried out with free culture of *Pichia stipitis* and immobilized *S. cerevisiae*. Pine wood was used in two forms; untreated (UPW) and treated (TPW). Fermentation experiments with free culture of both the yeasts were also carried out. The immobilization was carried out by co-culturing 48 h grown *S. cerevisiae* culture with PWC for 24 h with gentle shaking. The immobilization was confirmed by optical microscope and by the efficiency to ferment glucose quickly as compared to free culture. Initial experiments were carried out with 50 g/l mixture of glucose and xylose (1.5:1.0 ratio of glucose to xylose).

The efficiency of support for fermentation was evident by the fact that 63-70% of sugars were consumed in only 12 h and 84-93% fermentation efficiency was observed with almost complete utilization of both the sugars in 44 h. Doubling the sugar concentration (100 g/l) had little effects on fermentation efficiency but time required for complete sugar fermentation was increased to 78 h from 44 h. TPW worked out as a better support than UPW as the surface area of the former increased after the treatment to 1.5 times.

See more of this Session: Advances In Alternative BioEnergy
See more of this Group/Topical: Topical 9: Waste & Biomass Valorization

(40e) A Review of Biodiesel Production Via Reactive Distillation

Iman Noshadi[1], Nor Aishah Saidina Amin[1], Richard Parnas[2] and Seyed Masoom Khazraee[3], (1) Chemical Engineering, University Technology Malaysia, Johour, Malaysia, (2)Institute of Material Science, University of Connecticut, Storrs-Mansfield, CT, (3)Fars Engineering Research Center, Shiraz, Iran, Shiraz, Iran

Abstract— there are two main problems with continued use of fossil fuels, which meat about 80% of the world energy demand today. The first problem is that they are limited in amount and sooner or later will be depleted. The second problem is that fossil fuels are causing serious environmental problems. Since biodiesel is sustainable and combines high performance with environmental benefits it has become increasingly attractive as an alternative fuel. The conventional method for the production of biodiesel has several disadvantages: high alcohol demand due to shift equilibrium toward fatty esters, necessity of catalyst neutralization hence causing salt waste stream, high separation costs of fatty esters products from the reaction mixture, high production costs caused by complex processes including one or two reactors and several separation units. The reactive distillation (RD) process for the production of biodiesel overcomes these disadvantages. This paper attempts to review production of biodiesel by reactive distillation. The effect of various parameters feed ratio, type of applied catalyst, feed temperature, reboiler duty, type of column have also been summarized in this paper.

See more of this Session: Advances In Alternative BioEnergy
See more of this Group/Topical: Topical 9: Waste & Biomass Valorization

Lonza MicroReactors – Scale-up

EHRFELD Mikrotechnik BTS
A Bayer Technology Services Company

Preclinical → Clinical Phase I - III → Commercial Production

Lab Reactor	Plate Reactors A6 & A5	Plate Reactor A4
		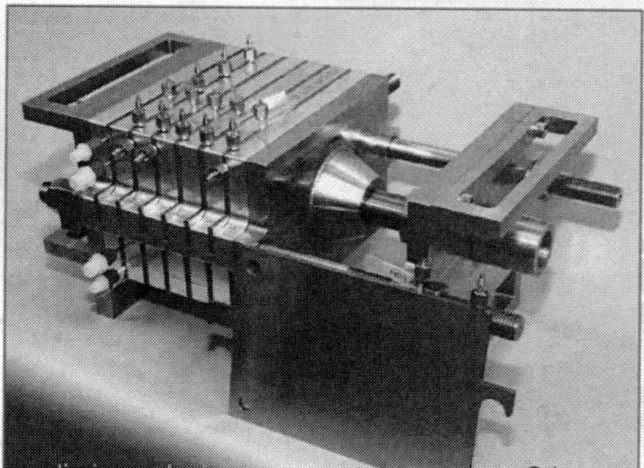
© Ehrfeld Mikrotechnik BTS	© Lonza	preliminary design © Lonza
1-50 g/min	30-150 & 100-300 g/min	200-600 g/min

Ehrfeld Microreactor Systems

EHRFELD Mikrotechnik BTS
A Bayer Technology Services Company

Pilot- and production scale apparatus for chemical industry

Modular MicroReaction Systems for laboratory use

Reactor systems for pharmaceutical industry

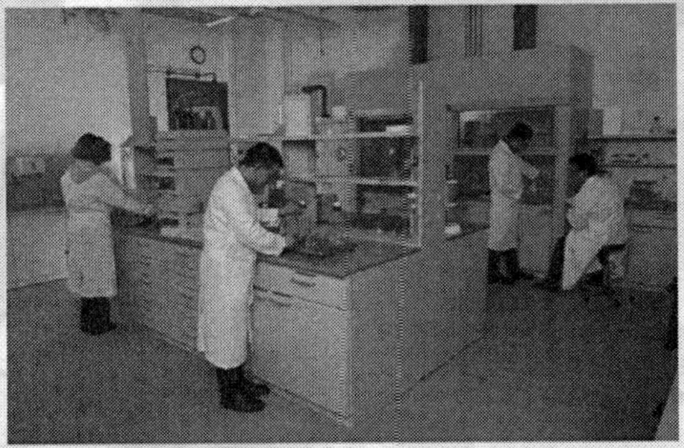

Feasibility studies, training, contract development

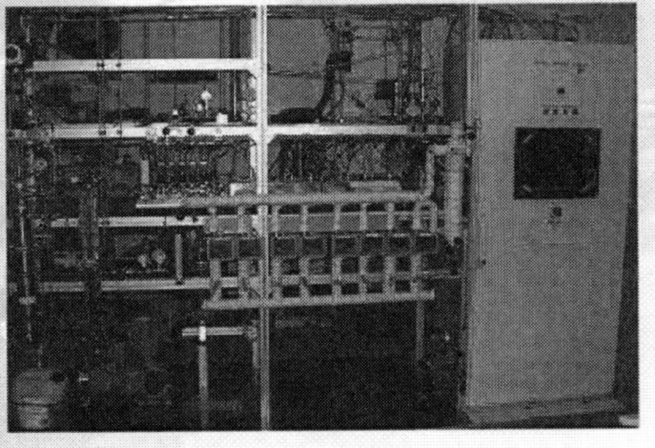

Customized plants

Paper 47b

PROCESS INTENSIFICATION OF MULTI-EFFECT EVAPORATION AND CRYSTALLIZATION PROCESSES

C. B. Panchal ** and Vyjayanthi Alagharu
E3Tec Service, LLC
Clarksville, MD 21029-1203
cpanchal@msn.com

Emmanuel Dada
ChemProcess Technologies, LLC
League City, TX 77573
emmanuel.dada@ymail.com

Chris Thomas
FMC Alkali Chemicals
Green River, WY 82935
Chris.C.Thomas@fmc.com

** To whom correspondence should be sent

Key Words: multi-effect, evaporation, crystallization, enhanced tubes, process intensification

Prepared for Presentation at the 2011 National Meeting, Chicago, IL March 14-17, 2011

Unpublished

PROCESS INTENSIFICATION OF MULTI-EFFECT EVAPORATION AND CRYSTALLIZATION PROCESSES

C.B. Panchal[a], Vyjayanthi Alagharu[a], Emmanuel Dada[b] and Chris Thomas[c]
[a] E3Tec Service LLC, Clarksville, MD 21029
[b] ChemProcess Technologies, LLC, League City, TX 77573
[c] FMC Alkali Chemicals, Green River, WY 82935

Abstract

Multi-effect evaporation and crystallization are key separation processes for concentrating and purifying inorganic and organic solid products. They are second only next to distillation in their consumption of energy with estimated energy consumption of greater than 660 trillion Btu/yr [ERDA, 1976], or the equivalent of 110 million barrels of oil per year. A small improvement in the energy efficiency can have major impact on overall energy savings and reducing the C-Footprint. The objective of this DOE SBIR Phase I project focused on two specific aims: 1) to evaluate enhanced-tube performance for process intensification; and 2) to characterize fouling propensity of enhanced tubes in multi-effect evaporation and crystallization. The process analysis was performed for the triple-effect and mechanical vapor recompression (MVR) crystallizers of the soda ash process. Once a benchmark analysis was performed, the impact of enhanced tubes was evaluated in terms of reduced power in MVR and reducing the primary-steam pressure in the triple effect crystallizer. The potential of energy efficiency in the MVR is direct and significant. The primary steam in triple-effect is generally derived from power generation; therefore, by reducing the steam pressure increases the power generation. Optimally designed enhanced tubes have shown to reduce fouling propensity in other applications. The interactive effects of wall temperature, wall shear stress, and secondary flows in the vicinity of the heated wall surface are the governing factors associated with fouling. Two types of enhanced tube were evaluated for applications in soda ash process. Applications of enhanced tubes in vertical or horizontal forced convective evaporator as well as falling-film evaporators were also evaluated.

Introduction

Process industries with heavy applications of multi-effect evaporation and crystallization include soda ash and caustic soda manufacturing, salt production, purification and separation of organics, black liquor concentration in the pulping process, recovery of distillers' dried grains with solubles (DDGS) in ethanol corn mills, production of ethylene and propylene glycol, agricultural chemicals, sugar production, fruit juice concentration, and various environmental treatments. The market potential of improved energy efficiency in multi-effect evaporation and crystallization is high. The production of soda ash and caustic soda are listed 9th and 10th in the US top 50 chemicals at 11,838 and 11,371 KTA (thousand tons per annum) [Energetics, 2000], and evaporated salt production is 4.3 KTA. Pulp, paper and paperboard mills account for 95% of energy use in the US paper and allied products industries, and 12% of total manufacturing energy use in the U.S [Nilsson et al., 1995]. Production of fuel ethanol is rapidly increasing in the U.S. to meet the national goal. Production is expected to increase from the current level of about 6 billion gallons per year to 36 billion gallons per year by 2020 [EERE, 2007], most of this production is expected to be based on corn and cellulosic feedstock. The net-energy value of corn-based ethanol can be achieved only when the energy credit for co-products, which requires multi-effect evaporators, is included [Shapouri et al.,

2002]. Therefore, it is essential to develop an energy-efficient evaporation process for product recovery.

Currently enhanced tubes are not used in multi-effect evaporators and crystallizers and monitoring-based fouling mitigation is not practiced. As a first impression, qualifying enhanced tubes in multi-effect evaporators and crystallizers seems an ordinary step and the industry should be able to pursue it without further developments. Response to this perception lies in understanding the complex multi-phase solid/liquid/vapor flows in evaporators and crystallizers. The industry is not likely to implement new enhanced tube geometry without validated thermal performance, effective and predictive fouling mitigation, and clearly defined economics in terms of Return on Investment (ROI).

Objectives and Specific Aims

The major objective of this SBIR project was to make fundamental changes in designing heat transfer equipment in multi-effect and MVR evaporation and crystallization processes that would have significant impact on process intensification; energy efficiency and productivity. In order to establish the applicability of enhanced tubes, this SBIR project focused on the following specific aims.

a. Optimal design of enhanced tubes in multi-effect and MVR evaporators and crystallizers;
b. Characterization of fouling propensity of enhanced tubes; and
c. Development of an innovative monitoring-based fouling mitigation system.

Application of Enhanced Tubes

This project focused on two types of enhancements; 1) Twisted Tubes®; and 2) spirally indented tubes with micro-fins on outside. They are ideally suited for single-phase flow, thin-film evaporation (rising or falling film) and condensation heat transfer in multi-effect evaporators and crystallizers. Twisted Tubes® (Figure 1) induce swirling flows that increases the thermal performance by more than 30% in single-phase flow and greater in two-phase flows, without corresponding increase in pressure drop. Spirally indented tubes, shown in Figure 2 enhance the thermal performance by interrupting the boundary layer and producing turbulent eddies in the vicinity of the wall without significant flow recirculation zones.

Figure 1 Enhanced Twisted Tubes® (Courtesy of Koch Heat Transfer Co.)

Selected tube inserts used in other process industries were also evaluated. A careful analysis of application of commercially available tube inserts showed that spirally twisted tapes or ribs and flow-induced oscillating wire inserts are viable candidates. Tube inserts can be installed without removing the tube bundle; therefore, it is a low-cost option for enhancing the thermal performance and reducing the fouling propensity. However, it is essential to optimize the geometry of tube inserts for multi-effect evaporator and crystallizer heat exchangers. The oscillating

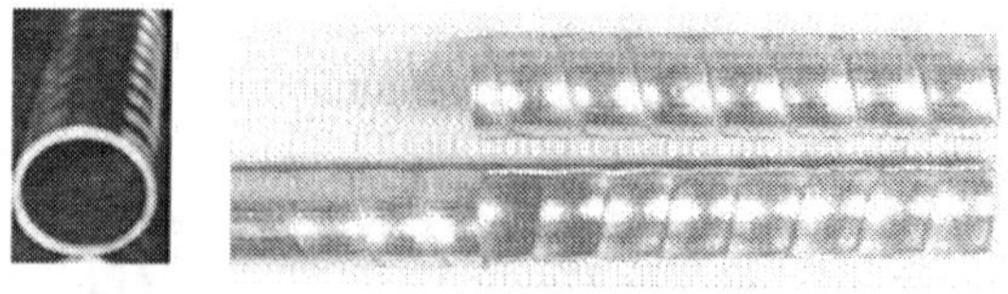

Figure 2 Spirally indented tubes

wire has shown to reduce fouling by generating secondary flows in the vicinity of the tube wall as well as disturbing the deposition mechanisms due to the oscillating movement of the wire. The major disadvantage of tubes inserts is that they cannot be readily removed for mechanical cleaning of heat exchanger tubes. Therefore, cleaning-in-place (CIP) methods are required for these inserts.

The major categories of heat-transfer equipment used in multi-effect evaporation and crystallization processes are represented by schematics in Figures 3 through 6. The caldaria evaporative crystallizer is generally used for salt production, granular compounds and sugar. Recirculation of solution is mostly caused by the thermo-siphon effects of two-phase flow inside tubes with added effects of recirculation impeller located in the center. Based on proven performance of twisted tube in vertical thermo-siphon reboilers in distillation processes [Panchal and Ljubicic, 2008], optimally sized *Twisted Tubes®* are ideally suited for calandria evaporative crystallizer.

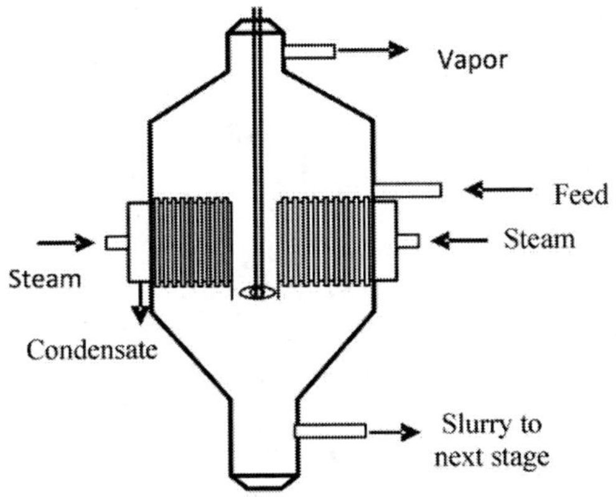

Figure 3 Calandria evaporative crystallizer

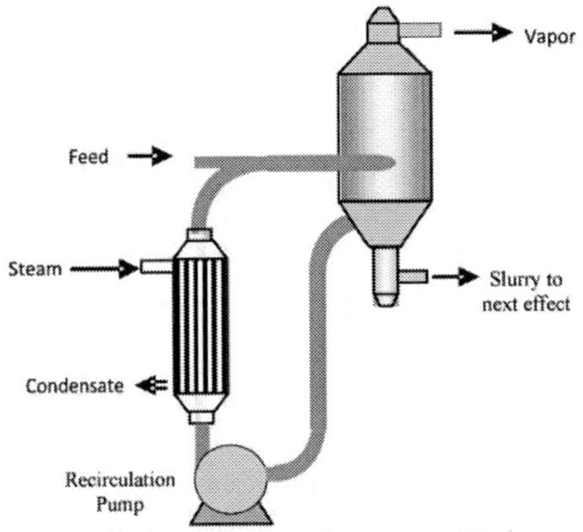

Figure 4 Forced-circulation evaporative crystallizer

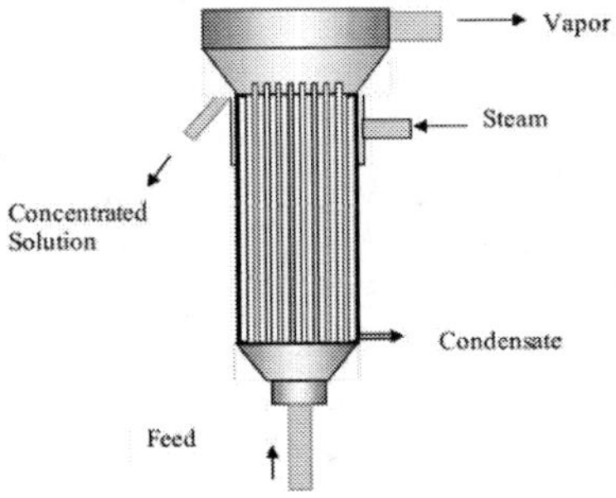

Figure 5 Vertical Rising-Film Evaporation

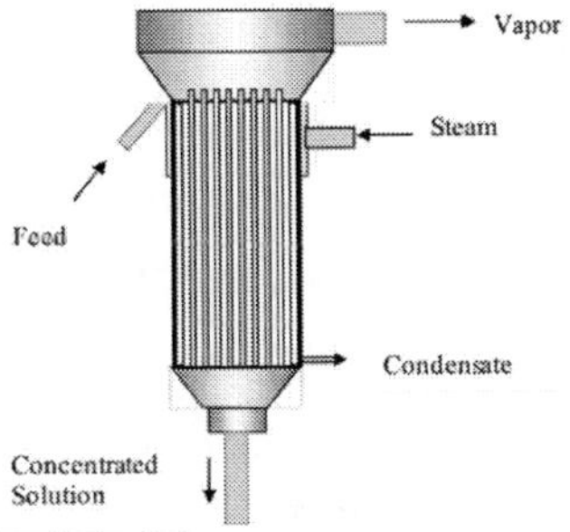

Figure 6 Vertical Falling-Film Evaporation

Forced-circulation evaporative crystallizers are widely used in various industrial processes, Figure 4. Evaporation can occur in the heat exchanger that would concentrate the solution. Alternatively, the solution is super-heated under static pressure, and the solution is allowed to flash producing low-pressure steam and crystal in the crystallizer. *Twisted Tubes®* are ideally suited for both modes of heat transfer; single-phase as well as phase-change. However, optimal size (twisting angle and pitch) would be different for single-phase and two-phase heat transfer modes. *Spirally indented tubes* should be considered for single-phase heat transfer only. Thin-film evaporation is commonly used for concentrating the solution without crystallization, such as caustic soda, food products, and in the paper pulping processes. Besides high thermal performance, thin film evaporation maintains the film temperature within a specified range for product quality and minimizing fouling deposition that can be caused by side reaction (specifically for food products) and crystallization (due to the presence of no-process elements) at higher wall temperatures. Twisted tubes should be considered for thin-film evaporation because swirling flow will maintain effective distribution of thin film on the wall; minimizing hot spots. In the MVR process, shown in Figure 7, vapor generated from the crystallizer or evaporator is compressed, which becomes the heat source. The enhanced thermal performance lowers the temperature difference between the compressed steam and the process fluid and thereby reduces the power consumption.

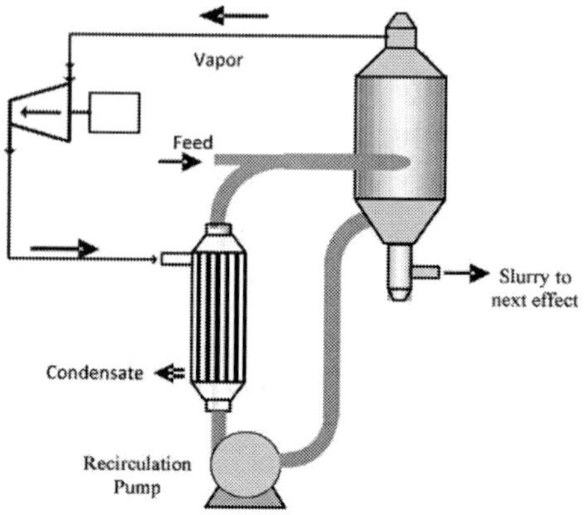

Figure 7 MVR Crystallizer/Evaporator

Steam Economy Analysis of Triple-Effect Soda Ash Process

A typical triple-effect soda ash process is shown in Figure 8. Primary steam is used in the 1st effect and steam generated in 1st and 2nd effects is used in the next effect as heat source. The soda ash solution is circulated at high flow rates to maintain temperature rise < 5 °C. The thermal performance of heat exchanger in each effect governs the temperature difference between the heat source (i.e. steam pressure) and process fluid. Poor thermal performance of heat exchangers would raise steam pressures in each stage, and thereby increase pressure of the primary steam. Source of the primary steam is generally the last stage of a power plant. Increase in steam pressure would adversely impact the power generation. According to EPRI guideline, 0.4" HgA increase in back-pressure of steam due to fouling of steam condenser corresponds to power penalty of about 0.5%.

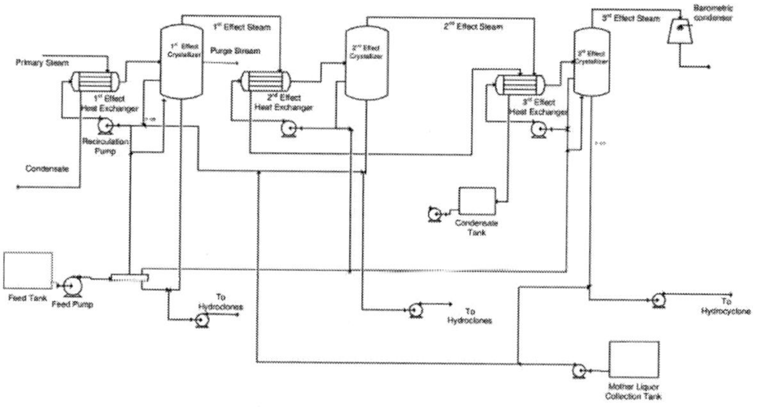

Figure 8 Process flow diagram of the soda ash triple-effect

The benchmark analysis performed by Aspen+® analysis followed by parametric analysis using SimuLink®. The analysis shows the benchmark steam economy is about 2.34. The steam economy could be even lower in normal operation due to poor thermal performance and fouling of heat transfer equipment. Log-mean temperature differences (LMTD) of the evaporative crystallizers vary between 10°C and 25°C, depending on recirculation rates, fouling, and the effectiveness of removal of non-condensable gases. By enhancing the thermal performance and reducing the fouling propensity of the three heat exchangers, LMTDs would be reduced with resulting effects on lowering the pressure of the primary steam from 257 kPa to a lower value of 152 kPa. This reduction in steam pressure would increase the power generation by 2% to 3%.

Performance Analysis of MVR

A typical MVR process diagram for soda ash is shown Figures 9. Aspen+® analysis was performed to establish the benchmark performance followed by SimuLink® analysis for parametric analysis. In addition, an Excel-based detailed heat transfer performance analysis was performed to determine an optimum enhancement. The benchmark power consumption of this case study is 69.7 kWh/metric ton (MT). The results show that by enhancing the heat transfer coefficient by 40%, which can be achieved by enhanced tubes described above, the power consumption can be reduced by about 9% by lowering the compression ratio. Further improvements would be achieved by reducing the effective fouling resistance and the effect of non-condensable gases. A conservative estimate of lowering the combined effects of fouling and non-condensable gases by one-half was used in this analysis. The resulting effects of heat transfer enhancement and lower effective fouling resistance are quite significant. The compressor power is reduced by 24% compared to benchmark power.

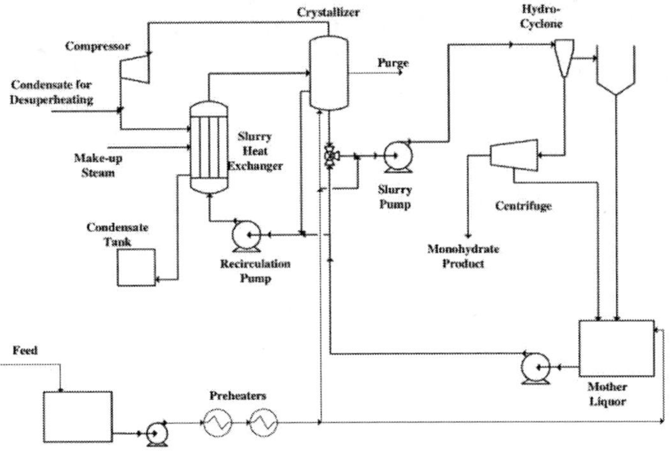

Figure 9 Process flow diagram of soda ash MVR

Fouling Characterization of Enhanced Tubes

In order to characterize fouling of enhanced tubes, it is important to understand the fouling mechanisms and interactive effects of governing mechanisms of precursor formation and subsequent physical processes associated with fouling deposition. Crystallization and particulate deposition are the two dominant fouling mechanisms in multi-effect evaporation and crystallization processes. Grace [1975] provided a survey of evaporator scaling of alkaline pulp industry, where crystallization fouling due to complex formation by non-process elements is a major concern. The fouling problem results from co-deposition of sodium carbonate (Na_2CO_3) and sodium sulfate (Na_2SO_4) [Schmidl and Frederick, 1998]. *With increased use of recycled water for achieving zero discharge, fouling caused by non-process elements are increasingly becoming a major concern in many industrial processes.*

Fouling in multi-effect evaporators and crystallizers can be modeled in terms of wall shear stress, wall temperature, and chemical/physical characteristics of foulant as shown by equation below developed for plain tubes.

$$\frac{dR_f}{dt} = R_f^*\left\{1 - \exp\left(-\frac{C\tau\lambda_f t}{\varphi}\right)\right\}$$

This equation is based on the assumption that foulant can be removed from the surface by wall shear stress. For crystallization fouling, increasing wall shear stress effectively enhances the heat transfer coefficient and hence the thickness of the thermal boundary layer and lowers the wall temperature closer to the process fluid temperature. The combined effects significantly reduce the fouling propensity. However, achieving higher wall shear stress, just by increasing the fluid velocity, would increase pressure drop and hence pumping power. With better understanding of the fouling mechanisms and fluid dynamics in the vicinity of enhanced tubes, the investigators can confidently conclude that by creating secondary flows would effectively reduce the fouling propensity. This is the basic concept of this SBIR project. Enhanced tubes create swirling flows in the vicinity of the wall surface that would have high effectiveness to reduce the fouling rate as expressed in equation above. Optimally designed spirally indented tubes would also induce secondary flows at micro scale in the vicinity of the wall, in addition to reducing the effective thermal boundary layer due to periodic interruptions. Panchal and research associates have developed the following threshold fouling correlation for organic fluid fouling involving chemical reaction in the boundary layer [Panchal et al. 1997]. This correlation is being used to evaluate threshold fouling conditions and design heat exchangers as well as characterizing fouling of enhanced tubes. These two fouling correlations can be integrated to develop a threshold fouling correlation for enhanced tube in evaporation and crystallization processes.

$$\frac{dR_f}{dt} = \alpha Re^{\beta} Pr^{-0.333} e^{\left(-\frac{E}{RT}\right)} - \gamma\tau$$

The influence of enhanced surfaces on transport processes, shear stress distribution, secondary flows in the vicinity of the heated surface, and wall temperature alter formation of fouling precursor and subsequent deposition. Crystallization fouling in evaporative crystallizers occurs in the thermal boundary layer. Therefore, the distribution of shear stress, secondary flows, and wall temperature have direct impact on the fouling characteristics of enhanced surfaces. The literature analysis clearly demonstrates advantageous characteristics of optimally enhanced tubes for mitigating fouling in single-phase as well as two-phase flow conditions. However, limited data are available to determine fouling characteristics of enhanced tubes in multi-effect evaporation and crystallization with free-flowing solids, which should be pursued using field-thermal monitoring unit as has been done for petroleum fouling [Panchal et al., 2004].

Conclusions

This SBIR project has shown ***techno-economic feasibility*** of ***enhanced tubes*** for applications in multi-effect evaporation and crystallization processes with potential of significant process intensification. The specific conclusions are as follows.

- Twisted Tubes® and spirally indented tubes, with or without micro-fins on outside for enhancing steam condensation, are potential candidates for multi-effect evaporation and crystallization processes.
- Aspen+® benchmark analysis combined with Excel-based thermal performance and SimuLink®-based parametric analyses provide effective tools to evaluate process intensification in multi-effect and MVR evaporation and crystallization processes.
- Improving the thermal performance of heat transfer equipment in the MVR process has high potential rate of return.
- Fouling monitoring units designed for the petroleum industry can be modified for multi-effect evaporation and crystallization processes.

Acknowledgement

This work was supported by the U.S. Department of Energy, the Office of Basic Science, Small Business Innovative Research (SBIR) Program.

References

ERDA, 1976, "Energy Conservation: Evaporation – A Prime Target for Energy Conservation," Energy Research and Development Administration Report, COO/2870-1.

Energetics, 2000, "Energy and Environmental Profiles of the U.S. Chemical Industry," Report prepared for the U.S. Department of Energy, Office of Industrial Technologies.

EERE, 2007, "Growing America's Energy Future," *The U.S. Department of Energy, EERE/ITP.*

Grace TM, 1975, "Survey of Evaporator Scaling of Alkaline Pulp Industry," Project 3234, Report 1, The Institute of Paper Chemistry, Appleton, WI.

Nilsson LJ, Larson ED, Gilbreath KR, and Gupta A, 1995, *Energy Efficiency and the Pulp & Paper Industry*, American Council for an Energy-Efficient Economy, Washington DC.

Panchal CB, Malone P, Stomeirowski S, and Cao T, "Mitigation of crude oil fouling by tube inserts and enhanced tubes," *Proceedings of the AIChE Spring Meeting*, New Orleans, LA, April 2004.

Panchal CB, Kuru WC, Liao CF, Ebert WA, and Palen JW, "Threshold conditions for crude oil fouling," *Understanding Heat Exchanger Fouling and Its Mitigation, Proceedings of the 1997 Engineering Foundation Conference*, Begell House, Inc., New York, NY 1997.

Panchal, CB, Malone, P., Stomeirowski, S., and Cao, T, 2004, "*Preheat Train Fouling: Effects of Process Parameters and Mitigation by Tube Inserts and Enhanced Tubes*," Proceedings of the AIChE Spring Meeting, New Orleans, La.

Panchal CB and Ljubicic B, 2008, "Design Approach for Mitigating Fouling of Reboilers and Feed/Effluent Heat Exchangers," Proceedings of the AIChE Meeting, New Orleans, LA.

Schmidl W, Frederick W; 1998, "Current Trends in Evaporator Fouling", Proceedings of International Conference on Chemical Recovery, TAPPI Press, Atlanta, GA.

Shapouri H, Duffield JA, Wang M, 2002, *The Energy Balance of Corn Ethanol*, The US Department of Agriculture Report No. 813.

(47c) Compact Heat Exchange Reactor for In Process Improvement

Zhijun Jia[1], **Steven J. Vallee**[1], Burtron H. Davis[2] and Dennis Sparks[2], (1)Chart Energy & Chemicals, La Crosse, WI, (2)Center for applied energy research, University of Kentucky, Lexington, KY

Chart has developed a next generation reactor that is very suitable for highly exothermic or endothermic reactions such as Fischer-Tropsch (FT) synthesis or Steam Methane Reforming (SMR). Chart's Compact Heat Exchange Reactor (CHER) is an advanced technology which combines heat exchanger, mixer, and reactor together in a compact structure. The reactor consists of multiple reaction and cooling layers alternately arranged. Each reaction layer contains multiple mini-channels each in the mm scale size range.

A study of FT synthesis using Chart's CHER was performed. Mini-channels in this reactor were packed with catalysts provided by Burt Davis's group at Kentucky University. The reactor was continuously running for three weeks at operating conditions of 220C and 15 bars. The tested reactor was made of aluminum which provides an excellent heat transfer characteristics. The results indicated that the compact heat exchange reactor provided unique features to improve the process efficiency and productivity. One of the significant benefits by using CHER in FT or SMR processes is improved temperature control which allows the reactor to operate efficiently at higher conversions than traditional tubular reactors.

See more of this Session: Process Intensification (I)
See more of this Group/Topical: Process Development Division

(47d) Membrane Extraction (pertraction) as a Tool for Process Intensification

Ikenna S. Ngene, Separation Technology, TNO Science and Industry, 2600 JA/Delft, Netherlands and Mark Roelands, Process Intensification, TNO Science and Industry, 2600 JA/Delft, Netherlands

Classical extraction is a well known technique used in the separation of components within the industry. Pertraction was developed as an extraction process which reduces the effect of emulsion formation resulting from mixing two solvent phases. In pertraction, a membrane is used to separate the two phases. Typically, the solute of interest is present in the aqueous phase and is transported into an organic phase across a hydrophobic membrane. Considering the independent mobility of the two phases, there is better control over optimization. With the use of fibers, the contact area for these pertraction processes can be significantly increased.

In terms of process intensification, there is presently a drive from batch to continuous processing in fine-chemicals industries. This drive offers advantages with respect to volumetric productivity, yield and selectivity, footprint and safety. Pilot studies of micro-structured reactors in a production environment are in progress. It makes, however, little sense to shift only the reactor to continuous operation and not the work up which would result in accumulation of the reaction mixture in storage vessels. The next logical step is the introduction of continuous intensified separation equipment that has a similar performance with respect to separation efficiency as continuous reactors regarding productivity.

This study is focused on the use of a flat sheet membrane contactor for pertraction with two different chemical applications tested. In the first part, an aqueous work up of a synthetic organic reaction mixture was carried out. This work up was aimed at the removal of spent organic acids and bases from the reaction mixture using pertraction. The work up was successfully carried out without emulsion formation and potentially with a smaller volume of washing water. The second application studied was the dehydration of a reaction mixture, with the objective of removing both dissolved and dispersed water formed. In the latter case the test system was an esterification reaction and dehydration into a hygroscopic solution. For both applications the mass transfer coefficient was determined in order to calculate the surface area of membrane required for an industrial module.

See more of this Session: Process Intensification (I)
See more of this Group/Topical: Process Development Division

New Optimization-Based Approach to Process Synthesis – towards the Full Integration of Process Design, Operation and Control

Nikola M. Nikačević [1], Adrie E.M. Huesman [1], Paul M.J. Van den Hof [1], Andrzej I. Stankiewicz [2]

[1] *Delft Center for Systems and Control, Delft University of Technology, Delft, The Netherlands,* n.nikacevic@tudelft.nl
[2] *Process & Energy Department, Delft University of Technology, Delft, The Netherlands*

Conceptual process design still follows a traditional procedure which consists of three sequential stages, as depicted in Figure 1. In the first stage, the operation mode is selected according to the particular process and manufacturing demands. The decision is strictly based on heuristic rules, for instance batch operation is preferable for small capacities and multi-products while continuous operation is recommended for bulk production of a single product. In the second stage, the process is designed according to the process' physical and chemical characteristics, desirable production rate and quality and related economical issues. Although in this step equipment dimensions are commonly calculated using mathematical models, this phase also heavily relies on heuristics. In most cases predefined process units for a specific application are used and a selection of units is completed according to this practical method. In the third phase, a control system is designed for the purpose of maintaining the specified product quality and quantity, as well as process safety. This procedure is well established in practice; it offers ready-made mathematical models and technical solutions. However, it does not allow the interaction of design, operation and control and therefore it is generally suboptimal from both economical and environmental perspectives.

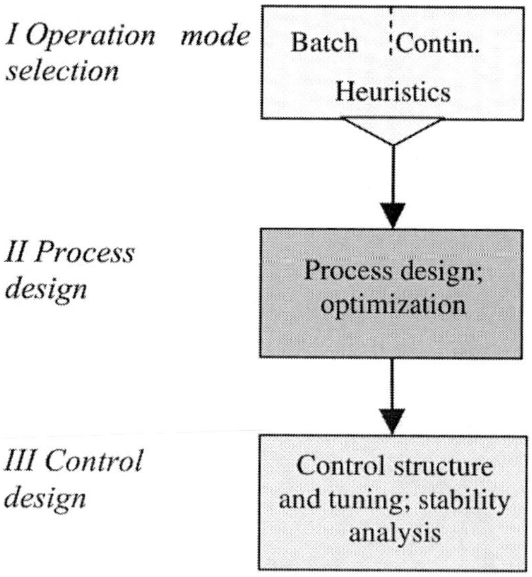

Figure 1. Traditional process synthesis

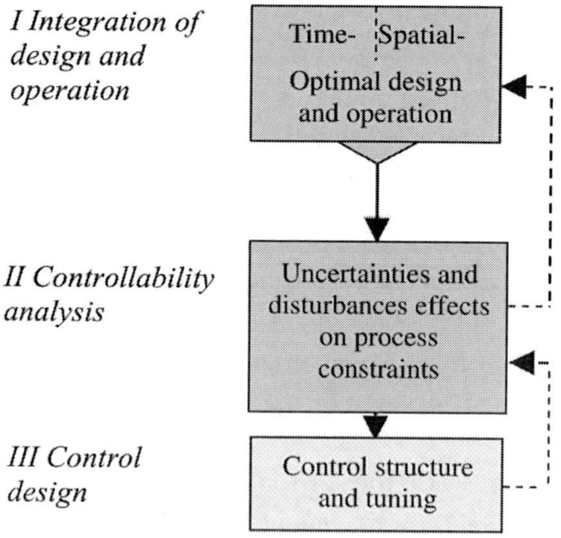

Figure 2. New optimization-based process synthesis

A new approach to chemical process synthesis is going to be presented and illustrated on an example. The concept is based on dynamic optimization and it also contains three stages as presented in Figure 2. In contrast to traditional approach this concept provides strong interactions between process operation, design and control. In the first stage, the complete integration of process design and operation should be attained by means of dynamic optimization, based on simple first principles mathematical models. Optimization objectives are based on economical modeling and they may include environmental issues. Optimization results should provide optimal process design and optimal operation. Proposed process synthesis concept is it is out of the limits of traditional unit operation approach and it is determined by exploration of phenomena and manipulation of fluxes and driving forces. It should be emphasized that the operation mode is not fixed *a priori* (as in Fig. 1) but is a result of the optimization. Therefore, this concept promotes the exploitation of process intensification principles and methods for process design. It also explores the possibilities for actuation improvement for optimal operation and control.

In the second stage (see Figure 2) controllability study for the optimal process solution (achieved in the first phase) is to be performed, by means of robust optimization techniques. In this analysis, the effects of model uncertainties, specifically physical parameters uncertainties, and process disturbances on the process constraints are to be examined. This analysis may lead to adjustment of actuation variables, in order to avoid violation of the constraints. In the third phase, the control system is conceptually designed, taking the controllability results into account. Its target is to follow the optimal operation path as effectively and robustly as possible. It is reasonable to assume that existing advanced control systems, like (nonlinear) model predictive control, or optimal control, will function adequately for the optimal designs and realization of the optimal operation. A closed loop information is transferred back to the stage two (dashed line) in order to perform closed-loop controllability analysis, which could eventually lead to additional actuation alteration (dashed line to the first stage). In this presentation, the first and most important stage of new process synthesis concept will be addressed, while the second stage will be tackled in the following paper.

The methodology for attaining the integration of process operation and design (stage one in Figure 2), could be generalized as follows. First, after a chemical problem is given (with known reaction mechanism and properties) an economic objective function is modeled, which includes the revenues for a production of desired product and costs for: a) raw materials b) investments, c) energy, d) environmental issues. The problem is then systematically analyzed to consider as much as (physically) possible structural and operational enhancements, in order to maximize fluxes and driving forces within the process. For that reasons various process intensification methods are considered. Formulation of a single unique mathematical model which incorporates different process structure options and ways of operation is not attainable at the present. Therefore, the proposed method implies the decomposition of a generic problem into two or more sub-problems to be subjected to an optimization (Figure 2). One type of general model will generate more spatially-oriented solutions, operated often in continuous regime, both steady and unsteady state. The second form of model will result in more timely-oriented solutions, operated more frequently in discontinues or semi-continuous mode. After the concurring problems are selected and models are set, all degrees of freedom (control variables) need to be listed and classified between time-variable or time-invariant. Apparent physical constrains are to be incorporated in the optimization set-ups, as usual. After dynamic optimizations (MIDO) are performed, the target function values will be compared for a selection of the more beneficial solution, which would further be treated in stages two and three.

Proposed process synthesis methodology, precisely its first phase, is illustrated on an industrially relevant example – methane steam reforming (MSR) accompanied with water gas shift reaction (WGS). The goals are to have: a) high conversion of methane; b) efficient separation of hydrogen; c) high driving forces for reaction and separation; d) low energy costs; e) low capital costs and f) compact design. To overcome the equilibrium limitations and achieve high conversion of methane, a functional integration of catalytic reactor with H_2 selective membrane is a very good process intensification option. A membrane will also provide immediate separation of a product from the undesired product and allow the use of lower temperatures. The alternative could be an integration of catalytic reaction with *in situ* adsorption; however this option was not consider at present, in order to have manageable and illustrative example. To attain high driving forces for the membrane separation, as well as for the reactions, consecutive order of process functions in one unit (reaction-separation-reaction-…) is considered, aside to fully integrated system. Without going into technical details, the spatially-distributed continuous systems could be realized as a tubular catalyst bed reactor containing a number of membrane-wall tubes for the H_2 permeate, if the process functions are integrated. The other, time-distributed class of systems could be designed as a rotating bed reactor with membrane-wall tubes. To minimize diffusion resistances, a structural packing with Ni catalyst coating will be assumed. It should have a large surface area, but high bed void fraction, similarly like foam type of packing.

Alongside mentioned structural and functional intensification, the actuation improvements should be examined systematically and rigorously within the optimization. The first reaction (MSR) is highly endothermic and requires high temperatures and the second reaction (WGS) is slightly exothermic and requires moderate temperatures. In order to attain optimal reaction temperature profile a spatially distributed heat supply will be considered and optimized. The optimal heat supply, with high energy efficiency could possibly be realized by using electric current as a direct heat sources in the catalyst. Alternatively, microwaves could be used for efficient and selective heat supply to catalyst. Further, optimal control in space and time with these alternative energy sources could be achieved effectively due to fast dynamic characteristics electromagnetic fields. For the continuous system, unsteady state (periodic) operation is also examined, as another possibility for operational improvement.

Economical target function comprises a maximal production of hydrogen with a minimal costs for catalytic and membrane surfaces and energy consumption. The dynamic models are simplified using common assumptions in order to attain convergence during the optimizations with a number of constraints and control variables. Altogether, six optimization problems have been derived and analyzed. For all analyzed systems the dynamic optimization results provide optimal design parameters and optimal operation, satisfying process constraints. Optimal design parameters include: number of parallel systems, number of membrane tubes per unit, diameters of units and membrane tubes, lengths, diameter of foam pore, etc. Optimal operation parameters are: inlet steam to methane ratio, inlet temperature and pressure, inlet flowrate of sweep gas, spatially distributed heat supply, etc. The base case system for the comparison of economic performance consists of one reaction unit followed by one separate membrane unit, similarly to a conventional system. Finally, the comparison of the objective functions (and the terms related to the profit and costs) between six analyzed systems suggests the overall optimal process solution. Beside that, the comparison of reaction rates and permeation rates profiles between several optimal process solutions, provide more insightful information and grounds for the selection of the optimal solution. In conclusion, the results demonstrate significant difference in economic and technological performance. This example demonstrates the new methodology well, and provides the basis for the further development of the suggested concept.

(62a) Setting Biorefinery Manufacturing Fundamentals to Produce a Portfolio of Commodity and Fine Chemicals

Alain A. Vertés[1], Masayuki Inui[1] and Hideaki Yukawa[2], (1)Molecular Microbiology and Biotechnology Group, Research Institute of Innovative Technology for the Earth (RITE), Kizugawa, Japan, (2)Molecular Microbiology and Biotechnology Group, RITE, Kizugawa, Japan

The chemurgy concept, that is, manufacturing industrial commodity and chemical products from organic raw materials such as agricultural biomass and lignocellulosic waste, predates the petrochemical industry. However, petrochemicals prevailed over fermentation chemicals given critical industrialization factors, comprising a low cost of the petroleum feedstock and a high process economic efficiency, as exemplified by the full utilization of products and by-products from petrochemical plants for the production of gasoline. These economies of scale and scope have played an important role in the booming industrial development at the beginning of the XXth century and were realized on a very large scale in World War II and the following decades[1].

These extremely efficient cost structures that have been derived in the petrochemical industry over a century are now both a blessing and a curse. The blessing is that these economies of scale and scope have made possible the manufacturing of cheap fuels and materials. The resulting abundant supply of chemicals for industry and transportation has enabled modern life and thus now appear to be indispensable. The curse is that efforts to displace and replace the petroleum industry by the chemurgy industry for alleviating global warming and fossil fuel supply threats meet a very high economic and resistance barrier[2]. This has resulted over the years in a detrimental delay in renewable chemistry technology investment, development, and deployment[3].

The biorefinery concept aims at recreating for agricultural feedstock cost structures that are similarly enabling as the ones at play in the petrochemical industry[4, 5]. This change is of course fundamental and will have far reaching consequences, as it integrates the agricultural value chain into the petrochemical value chain, thus colliding the transportation, polymer, and chemical value chains with other established value chains and notably of the food and feed industry[3]. A key difference nevertheless remains in that the higher oxygen content of biomass-based compounds makes possible to produce totally novel materials and chemicals with novels properties[6]. An enabling factor of the biorefinery vision is of course the availability of versatile, cost-effective, and robust industrial processes to manufacture an array of chemicals to serve various sectors including the fuel, chemical, industrial polymer, feed, food, cosmetic, and pharmaceutical industries. Notably, it is important for the biorefinery plant manager to be able to produce in addition to fuels, such as ethanol or isobutanol, higher value compounds for the chemical industry, such as ethyl-acetate or isobutanol.

We report here on a growth-arrested biotechnological process to manufacture ethanol, organic acids, and amino acids from mixed sugars, based on *Corynebacterium glutamicum*, an organism that has a long history of use to produce amino acids as feed and food supplements or cosmetic ingredients. The intrinsic design of the process is to decouple in a separate fermentor and a reactor biological catalyst-production and product-production phases[7] with cell recycling and using very high cell concentrations; it builds upon the intrinsic properties of this organism. *C. glutamicum* is a facultative anaerobic bacterium that grows in the absence of oxygen only in media containing a terminal electron acceptor such as nitrate[8]. *C. glutamicum* does not sporulate but remains metabolically active when placed under growth-arrested conditions[9, 10] and is not inhibited by common fermentation inhibitors present in typical lignocellulosic hydrolysates[11, 12]; moreover, it can catabolize a range of sugars in parallel with glucose[7, 13, 14]. A complete toolbox of molecular tools enables one to modulate at will its plastic genome and general metabolism[10, 15]. The RITE process presents the entire set of fundamental attributes necessary for reaching cost-effectiveness and enabling a rich product portfolio: simplicity, industrial robustness[16], raw material versatility (glucose, fructose, glucuronic acid, glucosamine, maltose, mannose, α-methyl-glucoside, ribose, sucrose, trehalose, arbutin, salicin, cellobiose, arabinose, xylose)[13, 14, 17, 18], product versatility (ethanol, organic acids, amino acids)[9, 16, 19, 20], reduced utilities requirement (reduced fermentation heat[12], no oxygenation requirement[9]), high conversion rates of carbohydrate feedstocks-to-product given cell recycling and the absence of growth requirements (e.g., 110 gl^{-1} or 40 $gl^{-1}h^{-1}$ D-lactate)[21], possibility to conduct continuous fermentations for long periods of time, dramatically reduced risks of contamination in the reaction vessel given the use of very high cellular concentrations, secreted product, no expensive reagent, portfolio of processes that enable the practitioner to manage the biorefinery plant in swiftly interchangeable processes or to capture synergies by implementing multiplex productions[22].

The stage is now set for this core technology to be translated into a biorefinery demonstration plant, as a first step towards full-scale commercialization.

References

1. Benninga, H. A history of lactic acid making. (Kluwer Academic Publishers, Dordrecht, The Netherlands; 1990).
2. Vertès, A.A., Inui, M. & Yukawa, H. Implementing biofuels on a global scale. Nat Biotechnol 24, 761-764 (2006).
3. Vertès, A.A., Qureshi, N., Blaschek, H.P. & Yukawa, H. (eds.) Biomass to biofuels: strategies for global industries. (Wiley, Chichester, UK; 2010).
4. Cherubini, F. The biorefinery concept: using biomass instead of oil for producing energy and chemicals. Energy Conversion Manag 51, 1412-1421 (2010).
5. Octave, S. & Thomas, D. Biorefinery: toward an industrial metabolism. Biochimie 91, 659-664 (2009).
6. Marquardt, W. et al. The biorenewables opportunity: toward next generation process and product systems. AIChE J 56, 2228-2235 (2010).
7. Vertès, A.A., Inui, M. & Yukawa, H. in 14th European Biomass Conference 1068-1071, Paris,

France; 2005).
8. Nishimura, T., Vertès, A.A., Shinoda, Y., Inui, M. & Yukawa, H. Anaerobic growth of Corynebacterium glutamicum using nitrate as a terminal electron acceptor. Appl Microbiol Biotechnol 75, 889-897 (2007).
9. Inui, M., Kawaguchi, H., Murakami, S., Vertès, A.A. & Yukawa, H. Metabolic engineering of Corynebacterium glutamicum for fuel ethanol production under oxygen-deprivation conditions. J Mol Microbiol Biotechnol 8, 243-254 (2004).
10. Yukawa, H., Inui, M. & Vertès, A.A. in Amino acid biosynthesis, Vol. 5. (ed. V.F. Wendisch) (Springer Verlag, Berlin, Germany; 2006).
11. Sakai, S. et al. Effect of lignocellulose-derived inhibitors on growth of and ethanol production by growth-arrested Corynebacterium glutamicum R. Appl Environ Microbiol 73, 2349-2353 (2007).
12. Inui, M. et al. in 28th Symposium on Biotechnology for Fuels and Chemicals, Nashville, TN, USA; 2006).
13. Kawaguchi, H., Sasaki, M., Vertes, A.A., Inui, M. & Yukawa, H. Engineering of an L-arabinose metabolic pathway in Corynebacterium glutamicum. Appl Microbiol Biotechnol 77, 1053-1062 (2008).
14. Sasaki, M., Jojima, T., Inui, M. & Yukawa, H. Simultaneous utilization of D-cellobiose, D-glucose, and D-xylose by recombinant Corynebacterium glutamicum under oxygen-deprived conditions. Appl Microbiol Biotechnol 81, 691-699 (2008).
15. Vertès, A.A., Inui, M. & Yukawa, H. Manipulating corynebacteria, from individual genes to chromosomes. Appl Environ Microbiol 71, 7633-7642 (2005).
16. Vertès, A.A., Inui, M. & Yukawa, H. Technological options for biolofical fuel ethanol. J Mol Microbiol Biotechnol 15, 16-30 (2008).
17. Yukawa, H. et al. Comparative analysis of the Corynebacterium glutamicum group and complete genome sequence of strain R. Microbiology 153, 1042-1058 (2007).
18. Kawaguchi, H., Vertès, A.A., Okino, S., Inui, M. & Yukawa, H. Engineering of a xylose metabolic pathway in Corynebacterium glutamicum. Appl Environ Microbiol 72, 3418-3428 (2006).
19. Inui, M. et al. Metabolic analysis of Corynebacterium glutamicum during lactate and succinate productions under oxygen deprivation conditions. J Mol Microbiol Biotechnol 7, 182-196 (2004).
20. Okino, S. et al. An efficient succinic acid production process in a metabolically engineered Corynebacterium glutamicum strain. Appl Microbiol Biotechnol 81, 459-464 (2008).
21. Okino, S., Suda, M., Fujikura, K., Inui, M. & Yukawa, H. Production of D-lactic acid by Corynebacterium glutamicum under oxygen deprivation. Appl Microbiol Biotechnol 78, 449-454 (2008).
22. Vertès, A.A., Inui, M. & Yukawa, H. Alternative technologies for biotechnological fuel ethanol manufacturing. J Chem Technol Biotechnol 82, 693-697 (2007).

See more of this Session: Developments in Biomass to Fuels
See more of this Group/Topical: Topical 9: Waste & Biomass Valorization

(62b) Evaluation of Pretreatment Methods In the Production of Ethanol From Cattail Leaves

Elijah Wreh, Biological Sciences, St. Cloud State Graduate School, Champlin, MN

Abstract

Previous research in this lab indicated that cattails are a potential source of biomass for the production of cellulosic ethanol since their carbohydrate composition is comparable to that of other plants being considered for biofuel production. To further test their viability, we tested various pretreatment methods on dried cattail leaves. Before polysaccharides in plants can be enzymatically hydrolyzed to fermentable sugars, the plant material must be pretreated to render the polysaccharides accessible to the enzymes. The purpose of this project has been to compare the efficiency of sulfuric acid and ammonia pretreatment methods in preparing cattail biomass for ethanol production. In this project, dried, powdered cattail leaves were pretreated either by autoclaving them with 2% sulfuric acid for one hour or by incubating them overnight at 40 C° in 15 % aqueous ammonia. Samples of the dried, pretreated solid were treated with cellulase and -glucosidase for 48 hours. To compare the efficiency of these pretreatment methods, glucose liberated in these samples was measured by a glucose oxidase assay. It was found that more glucose was recovered in the enzymatic hydrolysis (step two) than in the pretreatment step (step one.) In step two, more glucose was liberated from biomass pretreated with ammonia than from biomass pretreated with sulfuric acid. However, more glucose was recovered from in step one by sulfuric acid pretreatment. Overall 27.8 % of the starting biomass was recovered as glucose with ammonia pretreatment compared to 11.7 % for sulfuric acid pretreatment. Interestingly, 22.3% of the starting biomass was recovered as glucose when no pretreatment was used.

See more of this Session: Developments in Biomass to Fuels
See morc of this Group/Topical: Topical 9: Waste & Biomass Valorization

HEMICELLULOSE AND LIGNIN RECOVERY FOR AMMONIUM HYDROXIDE PRETREATMENT PROCESSES

*STEVEN R. SHERMAN, JACOB A. MORRIS, JOHN H. GOODELL**
Savannah River National Laboratory, Aiken, SC, USA
**Adrian College, Adrian, MI, USA*

Introduction

Pretreatment is a step performed in fermentation-based processes that convert lignocellulosic materials into fuels and chemicals. During pretreatment, size-reduced (granulated, chopped, ground, etc.) biomass is subject to physical or chemical steps to make it more amenable to hydrolysis into sugars [1]. While physical pretreatment methods operate by creating more biomass surface area on a microscopic level, chemical pretreatment operates by removing or disrupting the lignin structure that hinders access to cellulosic fibers in the biomass.

Since 2008, Savannah River National Laboratory (SRNL) and Clemson University have been funded by the U.S. Department of Energy, Office of Energy Efficiency and Renewable Energy (DOE-EERE), to develop a fermentation-based process for converting switchgrass, sorghum, and wood into ethanol and other biofuels. The process backbone consists of the following steps: size reduction, pre-treatment, hydrolysis, fermentation, and product purification. The Soaking in Aqueous Ammonia (SAA) pretreatment method [2–4] was initially selected for this work due to prior experience with the method, and it has since proven effective at pretreating these feedstock materials. Application of this method dissolves between 40-60% of lignin, $<$ 20% of hemicellulose, and $<$ 5% of the cellulose from the pretreated biomass. Although the amount of lignin dissolved by this method may be less than can be achieved using other methods, this amount of lignin dissolution has proven sufficient to allow for greater than 80% hydrolysis of pretreated biomass into into sugars when commercially available cellulase mixtures are used to perform the hydrolysis.

As this work proceeds, we have begun to appreciate that the economics of the fermentation-based process may be improved by generating multiple products from the same feed material. During the pretreatment step, lignin is removed from the biomass. This lignin may be recovered from the pretreatment fluid and used as a fuel or a feedstock material higher-value chemicals. Following pretreatment, cellulose is hydrolyzed into glucose (a C-6 sugar), and hemicellulose is hydrolyzed into xylose (a C-5 sugar) and other sugars. In the fermentor, C5-sugars such as xylose are not easily metabolized by natural yeast. As a result, process throughput is reduced and the break-even price of the ethanol product is increased. Although C5-sugar utilization could be improved by using a genetically modified yeast or alternative organism, ethanol has a relatively low market price, and the C5-sugars might be better used to make products that have a higher market value than ethanol.

SRNL recently developed a process for separating and isolating lignin from ammonium hydroxide pretreatment fluids [5], which allows for the generation of a lignin side stream. Some hemicellulose is also extracted during ammonium hydroxide pretreatment, but it is not recovered by the lignin recovery process. Hemicellulose is better extracted using a separate process step, and hemicellulose extraction methods are described in the literature that might be used to recover hemicellulose as a distinct product [4, 6]. When performed before the hydrolysis step, both extraction steps are categorized as pretreatment.

In this work, the performance of the lignin extraction step and the hemicellulose extraction step are examined separately, and then in combination. When both extractions are performed in sequence, the order of the steps is alternated and compared, so that a recommended sequence can be discovered. The lignin extraction is performed using the SRNL method cited previously. The hemicellulose extraction is performed using a method adapted from [6] that uses $Ca(OH)_2$ solution as an extracting agent. Due to lack of chemical analytical data, the chemical composition of the lignin and hemicellulose samples collected during the course of this work is unknown. Instead, sample color and appearance are used as qualitative indicators of product purity in the tests.

Description of Lignin Recovery Method

Lignin is recovered from the ammonium hydroxide pretreatment fluid by the following procedure, and is described more fully in [5].

- L1: The pretreatment fluid is separated from the pretreated biomass material using filtration.
- L2: The ammonium hydroxide pretreatment solution containing dissolved lignin and other materials is heated to the boiling point in order to evaporate ammonia from solution. Ammonia-rich vapor is allowed to evaporate from the solution until the boiling point of the solution reaches 96-98°C at 101.1 kPa.
- L3: While hot, the solution is titrated with 10% sulfuric acid solution until the color of the solution changes from black to brown and the pH of the solution falls to between 2 and 3. The color change indicates the precipitation of a colloid of lignin particles.
- L4: The titrated solution is heated again to the boiling point, and water vapor is allowed to evaporate from the titrated solution until the volume of liquid has fallen to approximately half the starting volume and a "skin" of brown solid begins to form on the sides and top of the liquid. The formation of the "skin" indicates coagulation of colloidal lignin particles into larger particles that are separable from solution by gravity or filtration.
- L5: Allow the mixture to cool, and separate the lignin particles from the liquid by high-speed centrifugation or filtration using a coarse ($>25\ \mu m$) filter and then a fine ($>4\ \mu m$) filter.
- L6: Dry the recovered solids in the open air, or in a vacuum oven at a temperature less than 65°C to avoid conglomeration and glazing of the recovered solid.

Description of Hemicellulose Recovery Method

A hemicellulose extraction method adapted from [6] was used to extract and isolate hemicellulose from untreated and pretreated biomass in this work. The extraction steps are provided below. This method was selected for this work because it was well described in the literature, was free to use because the applicable patents had expired, and could be performed using the same equipment as was used for performing lignin extractions.

- H1: Size-reduced biomass (untreated or pretreated) is combined with 0.25 wt% $Ca(OH)_2$ solution in a 10:1 (volume-to-mass) ratio, and the slurry is mixed and heated for 2 hours at 80°C.

- H2: The slurry is allowed to cool, and then the liquid is separated from the solid material by filtration.
- H3: The recovered filtrate is titrated using 10-20% sulfuric acid until the pH of the filtrate is between 2 and 3. The solution may appear yellow-tinted with a fine suspension of white material after titrating it with sulfuric acid.
- H4: The titrated fluid is mixed with methanol in a volumetric ratio of 1.25:1 (volume-to-volume). This causes hemicellulose to precipitate out of solution. The appearance of the solution changes from pale yellow to milky white.
- H5: The precipitated solid may be allowed to settle and the liquid decanted, or the solid-liquid separation process may be accelerated by high-speed centrifugation or filtration.
- H6: The recovered solid, which is white, is allowed to dry in the open air or in a drying oven at a low temperature to avoid melting or caramelizing the recovered hemicellulose.

Results and Discussion

All tests were performed using 50 g samples of biomass as the starting amount of material. The amount of pretreatment solutions used to extract lignin and hemicellulose from the biomass materials were then adjusted to match the mass of the starting material.

Figures 1 and 2 show the percentage of original lignin recovered from previously untreated biomass. Four different feedstock materials were tested: switchgrass, sweet sorghum, sweetgum, and loblolly pine. The percentage of original lignin recovered is defined as the mass of lignin recovered as a solid lignin product divided by the estimated amount of lignin originally present in the biomass, based on characteristic compositional data provided in [8]. According to Figure 1, the amount of lignin recovered varies with the type of biomass when the pretreatment ratio (volume:mass) is 8:1. At an 8:1 pretreatment ratio, the lignin extraction method worked best on switchgrass, and was the least effective on loblolly pine. When the pretreatment ratio was increased from 8:1 to 16:1, the performance of the lignin extraction method on sweet sorghum and loblolly pine was improved, and, for loblolly pine, became statistically indistinguishable from the results obtained for switchgrass (see Figure 2). The convergence of results when the pretreatment ratio was increased from 8:1 to 16:1 suggests that ammonium hydroxide pretreatment, rather than the lignin recovery process, is responsible for the poorer lignin recovery from sweet sorghum, sweetgum, and loblolly pine. The ammonium hydroxide pretreatment process may be affected by a lignin solubility limit, the composition of the biomass, or, more likely, mass transfer limitations that are alleviated at higher liquid-to-solid ratios. The solid material recovered from all of the biomass samples using the lignin recovery procedure was dark brown and powdery, and matched the appearance and consistency of samples that were chemically analyzed and identified as lignin in previous work [5].

The fraction of original hemicellulose recovered from previously untreated samples of switchgrass and sweet gum is shown in Figure 3. In the figure, only about 20% of the original hemicellulose content in the switchgrass samples was recovered as a solid product, while more than 40% of the original hemicellulose content was recovered from samples of sweet gum. Since the same extraction and hemicellulose recovery procedure was used for both materials, it is likely that the differences in the results are due to mass transfer limitations during the extraction. No additional work was performed to

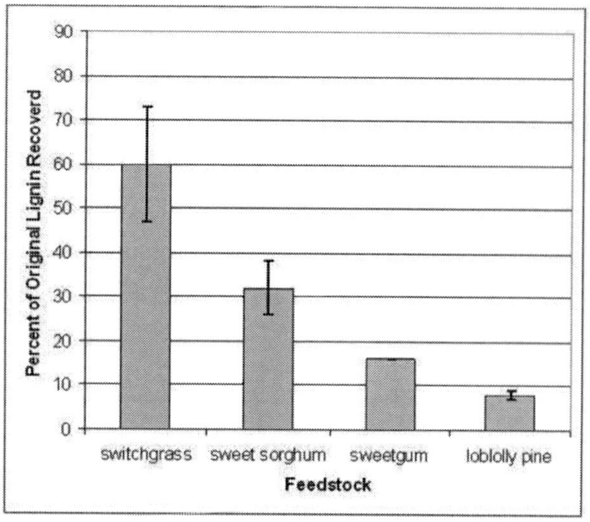

Figure 1: Lignin recovery varies with feedstock at an 8:1 pretreatment ratio.

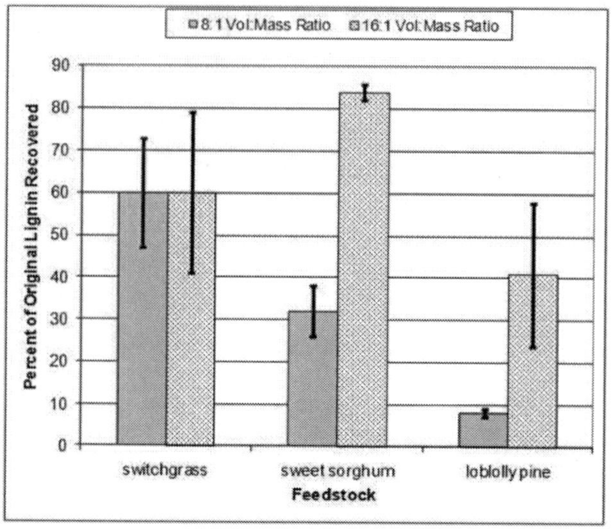

Figure 2: Lignin recovery is improved at a 16:1 pretreatment ratio.

study this question, and more work remains to be done if this hemicellulose extraction method is to be improved and optimized for the biomass feed materials of interest. The hemicellulose recovered from switchgrass had a dull, white, semi-crystalline appearance, while the hemicellulose recovered from sweet gum was bright white and powdery. Both types of hemicellulose turned light brown in color when heated in a drying oven.

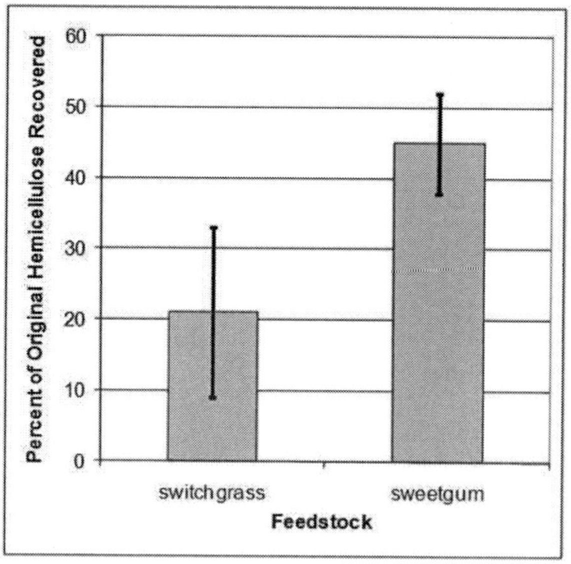

Figure 3: Hemicellulose recovery varies with feedstock.

The switchgrass and sweet gum samples from which hemicellulose had been extracted were then subjected to ammonium hydroxide pretreatment, and lignin was recovered from the ammonium hydroxide pretreatment fluid. The percent of original lignin recovered for these samples is shown in Figure 4. The percentage of original lignin recovered was lower for switchgrass from which hemicellulose had been extracted than for previously untreated switchgrass. For sweetgum, there was no distinguishable difference in the percentage of original lignin recovered, though it is difficult to assign a clear trend for sweetgum when the error bars on the data are so wide.

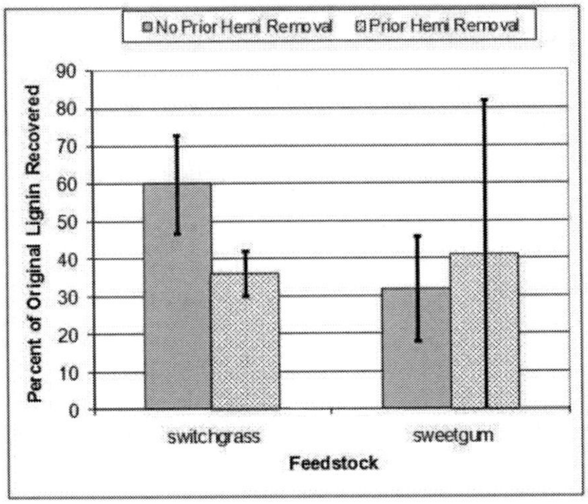

Figure 4: Variation in lignin recovery after some hemicellulose has been removed from the biomass.

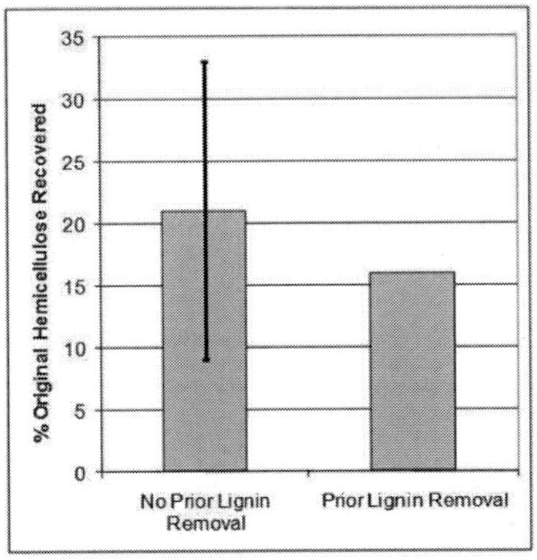

Figure 5: Comparison of hemicellulose extraction results on untreated and NH_4OH pretreated switchgrass.

Figure 5 shows a comparison of the percentage of hemicellulose recovered from untreated switchgrass, and switchgrass that had been pretreated with ammonium hydroxide. Although the chart shows a lower percentage of hemicellulose recovered from switchgrass that had been pretreated, statistically there is no difference between the two cases, and it appear that hemicellulose removal is not affected by ammonium hydroxide pretreatment, at least for the experimental conditions tested here.

Conclusions

The lignin extraction procedure and hemicellulose extraction procedure were shown to be capable of recovering lignin and hemicellulose as separate products from a range of biomass materials (i.e., switchgrass, sweet sorghum, sweetgum, and loblolly pine). The amount of lignin recovered from each type of biomass varied, and it appears that the variation in the amount of lignin recovered was due to mass transfer problems during pretreatment, rather than due to the lignin recovery procedure. The percentage of original hemicellulose recovered from sweetgum was higher than was recovered from switchgrass, which shows that the hemicellulose extraction procedure is affected by biomass type and perhaps mass transfer limitations. The amount of hemicellulose recovered from switchgrass that had been pretreated with ammonium hydroxide was lower than the amount recovered from untreated switchgrass, though this trend was not seen with sweetgum. The amount of lignin recovered from switchgrass that had previously been pretreated with $Ca(OH)_2$ solution was statistically similar to the amount of lignin recovered from untreated switchgrass.

With no clear trends in the data indicating an ideal sequence of extraction steps, it appears that the order in which the hemicellulose and lignin extraction steps is not important. Although there may be some amount of hemicellulose that is extracted along with lignin during ammonium hydroxide pretreatment, apparently the amount of hemicellulose lost from the biomass during the lignin extraction step is small enough that it does not significantly affect the performance of the subsequent hemicellulose extraction step. The reverse sequence, performance of the hemicellulose extraction step before extracting lignin with ammonium hydroxide, shows no effect on the amount of lignin recovered from switchgrass, and so is also a valid extraction sequence.

The variation in the data are wide. More experiments with repeated trials are needed to generate better statistical information before strong conclusions can be drawn on the proper sequencing of the extraction steps. Also, it would be helpful to collect chemical compositional data on the pretreated biomass and recovered products to discern the effects of varying the sequence of extractions on biomass composition and product purity.

References

[1] Y. Sun, J. Cheng, "Hydrolysis of lignocellulosic material for ethanol production: a review", *Bioresource Technology* (2002), 83:1-11.

[2] T.H. Kim, F. Taylor, K.B. Hicks, "Bioethanol production from barley hull using SAA (soaking in aqueous ammonia) pretreatment", *Bioresource Technology* (2008), 99:5694-5702.

[3] T.H. Kim, Y.Y. Lee, "Pretreatment of corn stover by soaking in aqueous ammonia at moderate temperatures", *Applied Biochemistry and Biotechnology* (2007), 137-140:81-92.

[4] T.H. Kim, Y.Y. Lee, "Fractionation of corn stover by hot-water and aqueous ammonia treatment", *Bioresource Technology* (2006), 97:224-232.

[5] S.R. Sherman, J.J. Goodell, C.E. Milliken, J.A. Morris, M.B. Gorensek, "A New Process Developed for Separation of Lignin from Ammonium Hydroxide Pretreatment Fluids", accepted for publication in *Environmental Progress and Sustainable Energy*, December 2010.

[6] M.W. Rutenberg, W. Herbst, "Process for Extraction of Hemicellulose", U.S. Patent 2,801,955, Patented August 6, 1957.

[7] W. Shu-Lai Mok, M.J. Antel Jr., "Uncatalyzed Solvolysis of Whole Biomass Hemicellulose by Hot Compressed Liquid Water", *Industrial Engineering Chemistry Research* (1992), 31:1157-1161.

[8] W.J. Frederick Jr., S.J. Lien, C.E. Courchene, N.A. DeMartini, A.J. Ragauskas, K. Iisa, "Production of ethanol from carbohydrates from loblolly pine: A technical and economic assessment", *Bioresource Technology* (2008), 99:5051-5057.

(62d) Accelerated Hydrolysis of Starch Using Cationic Polyacrylamide for Production of Biofuels

Kendra E. Maxwell, School of Chemical & Biomolecular Engineering Georgia Tech, Georgia Institute of Technology, Atlanta, GA and Sujit Banerjee, School of Chemical & Biomolecular Engineering, Georgia Institute of Technology, Atlanta, GA

Improving the efficiency of the hydrolysis of starch to glucose is key to reducing energy use during the conversion of starch to biofuels. We have found that the addition of positively charged polyacrylamide increases the conversion rate of starch to glucose. The aim of this study is to determine the major factors which affect the mechanism so that optimal production rates are achieved in the presence of the cationic polyelectrolytes. A combination of analytical methods including dynamical light scattering, optical imaging, and uv-vis spectroscopy was used to study polyelectrolyte, starch, and enzyme interaction as a function of process conditions crucial to starch hydrolysis such as composition and temperature. The results indicate that interactions caused by electrostatic forces largely drive the increase in hydrolysis rate. These and related results will be discussed in this presentation.

See more of this Session: Developments in Biomass to Fuels
See more of this Group/Topical: Topical 9: Waste & Biomass Valorization

(62e) Production of Bio-Fuels From Giant Miscanthus

Sanjeev K. Gajjela, Brain Mitchell, Qi Li, El barbary M. Hassan and Philip H. Steele, Forest Products, Mississippi State University, Mississippi State, MS

The energy crop giant miscanthus was pyrolyzed in a laboratory-scale auger reactor at Mississippi State University (MSU). The influence of pyrolysis temperature and biomass pyrolysis time on bio-oil yields is investigated. Pyrolysis temperature of giant miscanthus was varied between 375-450 °C. The highest bio-oil yield of 60 wt% was obtained at 425 °C. Best yield was also a function of biomass feed rate during pyrolysis. We found that maximum yield was attained for an approximate doubling of feedstock feed rate for giant miscanthus as compared to that for pine wood. The reason for this is likely due to differences in particle thickness with wood particles having approximately double or more the thickness of pine wood particles. This apparently allowed the giant miscanthus feedstock particles to pyrolyzed fully at a higher rate. When fed at a the same rate as for pine particles over pyrolysis of the miscanthus occurred as evidenced by lower yields and higher water content of the resulting bio-oil.

Further, raw bio-oil from giant miscanthus in the presence of hydrotreating/hydrocracking catalyst with high temperature and high hydrogen pressure yields hydrocarbons and water. MSU has developed effective hydrotreating (HT) and hydrocracking (HC) catalysts that have successfully produced a hydrocarbon mixture from pine-based bio-oil. These catalysts were applied to the hydroprocessing of giant miscanthus bio-oil. Giant miscanthus bio-oil demonstrated some differences in response to the first-stage HT and second-stage HC treatments as compared to pine bio-oil. These differences will be described in our paper. The physical and chemical properties of giant miscanthus bio-oil and upgraded bio-oil are determined. The fuel-value characteristics of the hydrocarbons produced by two-stage HT and HC will be fully described by analysis by GC/MS and FTIR. Simulated distillation will be performed on the hydrocarbons to describe the boiling range distribution of the hydrocarbon weights produced with comparison of these weights to those of petroleum fuels in the same molecular weight classes.

See more of this Session: Developments in Biomass to Fuels
See more of this Group/Topical: Topical 9: Waste & Biomass Valorization

DEVELOPMENT OF A DETAILED REACTION KINETIC MODEL FOR CELLULOSE FAST PYROLYSIS

Vinu Ravikrishnan and Linda J. Broadbelt, Chemical and Biological Engineering, Northwestern University, Evanston, IL

Fast pyrolysis is one of the potential strategies to convert whole biomass to sustainable transportation fuels. Fast pyrolysis involves heating the biomass in an inert atmosphere to moderate temperatures (400 – 500 °C) in a very short period of time (< 1 s). This leads to a complex network of competing reactions, which results in the formation of bio-oil, non-condensable gaseous species, and solid char. Bio-oil is a mixture of anhydro sugars, phenolic compounds, furan derivatives and low molecular weight oxygenated compounds. The development of a detailed reaction kinetic model to predict the yields of the various pyrolysis products is of utmost importance to understand the relative contributions of the various reaction pathways, and the effect of process parameters like temperature, pressure, and mineral content of biomass.

The existing kinetic models of biomass pyrolysis are global, as they lump the various product species into different fractions like char, condensable and non-condensable gaseous species, thus ignoring the individual identity of the reaction steps that lead to the various products. We have developed a comprehensive mechanistic model for the fast pyrolysis of cellulose, which incorporates the individual steps for the formation of important products like levoglucosan, glycolaldehyde, 5-hydroxymethyl furfural, furfural and formic acid. The underlying mechanism involves the scission of the cellulose chain both at random mid-position and at the chain ends. The heterolytic glucosidic bond scission and the retro-Diels-Alder reactions occur at the mid-chain, while the retro-aldol reaction, 1,2-dehydration, 1,3-dehydration, hydrolysis, and cyclic Grob fragmentation reactions occur at the chain end. The kinetic rate coefficients for most of the above reaction steps were obtained from the literature, based on either experimentally determined Arrhenius parameters for cellulose and model compounds, or theoretically computed values of activation energies using quantum calculations. A computational framework based on continuous distribution kinetics was constructed to solve the kinetic model and predict the time evolution of the product yields. The predictions are compared with experimental data from different sources reported in the literature. The promising results of this study and the applicability of the model for complex biomass systems will be addressed in this presentation.

Production of Dimethyl Carbonate via Reactive Distillation Process

Emmanuel A. Dada[1*], Chandrakant B. Panchal[2], and Richard D. Doctor[2] (1) ChemProcess Technologies (CPT), LLC, League City, TX 77573, (2) Energy System Division, Argonne National Laboratory, Argonne, IL 60439.
*: Presenter and corresponding author, emmanuel.dada@ymail.com

Abstract:

Dimethyl Carbonate (DMC) is an environmentally benign and biodegradable product. Because of its high oxygen content, DMC is an idea additive for gasoline to replace methyl tert-butyl ether (MTBE). DMC has many applications in the synthesis of polycarbonates, production of methyl isocyanate and as a replacement of halogenated solvents. It is extensively used as a methylating and carbonylating agent. Dimethyl Carbonate has recently becomes an important solvent in the lithium ion battery solutions and other electrochemical and coating sectors.

The early commercial production of DMC via phosgenation of methanol has many draw backs especially the toxicity of phosgene and is being phased out and replaced by other methods. Among these new routes of producing DMC are the oxidative carbonylation of methanol involving CO, transesterification using propylene carbonate or ethylene carbonate to co-produce DMC and propylene glycol or ethylene glycol, and methanolysis of urea. The merits of each of the new routes of producing DMC will be presented especially the systematic choice of entrainers in the extractive distillation step to separate the DMC-Methanol azeotrope common to all the three routes.

Introduction:

DMC has received significant research efforts in recent years due to its versatile chemical properties and commercial applications. It was first manufactured in the 1910s from methanol and phosgene. This process has many draw backs especially the toxicity of phosgene and is being phased out and replaced by other methods. One of the most important non-phosgene routes which is of the focus for our project is the transesterification of ethylene carbonate with methanol to produce dimethyl carbonate (DMC) carried out in a reactive distillation column. The common separation issue to most of the routes is the separation of the azeotrope mixture of methanol and dimethyl carbonate to make high purity DMC product. This issue is being addressed in our project with a focus to use a heat-integrated reactive distillation method couple with a heat-integrated extractive distillation for high purity DMC product that will be an economically promising and energy savings route.

Design of the Extractive Distillation:

The choice of the best entrainers that is environmentally acceptable, safe, and is relatively cheap is very critical for this process. In this respect we have considered the residue curve maps and the influence of some entrainers on the activity coefficients (related to the separation factor-relative volatility) of the Methanol-DMC azeotrope mixture as shown below figures and table 1. The maps were generated using ICAS software version 13 (2010). Details of the entrainers' selection process will be presented.

Table1: Relative Volatility of Methanol to Dimethyl Carbonate at Infinite Dilution

	Azeotrope Feed plus 50% Aniline, Fig 4	Azeotrope Feed plus 50% Phenol, Fig 1	Azeotrope Feed plus 80% Aniline, Fig 5	Azeotrope Feed plus 80% Phenol, Fig 2	0% Aniline, 0% Phenol, Figure 3
Activity Coeffcient of Methanol at infinite dilution, GammaI	2.3	1.91	2.1	1.1	2.43
Activity coeffcient of DMC at infinite dilution, GammaJ	0.93	0.72	0.81	0.3	0.99
Selectivity of I,J =GammaI/GammaJ	2.47	2.65	2.59	3.67	2.45
Relative Volatility = Selectivity of I,J * Ratio of Pure Comp. Vap Pressure (I to J)	6.35	6.81	6.65	9.41	6.28

Table 2: List of Potential Entrainers for Dimethyl Carbonate-Methanol Mixture

Name	Normal Boiling Point, K	Normal Melting Point K	Dipole Moment Debye	Solubility Parameter (MPa)^0.5	Reference
Aniline C_6H_7N	457.15	267.13	1.5288	24.1171	Hsu, K. et al (2010)
Phenol C_6H_6O	454.99	314.06	1.4508	24.6326	Wang, S.J. et al (2010)
Ethylene Glycol $C_2H_6O_2$	470.45	260.15	2.3082	33.7033	Hsu, K. et al (2010)
Ethylene Carbonate $C_3H_4O_3$	511.15	309.55	4.5095	30.0352	Himmele, W. et al (1979)
Diethyl Oxalate $C_6H_{10}O_4$	458.85	232.55	2.4897	20.1533	Ryu and Gelbein (2002)
Chlorobenzene C_6H_5Cl	404.87	227.95	1.6906	19.3453	Li and Zhang (2006)
Benzonitrile C_7H_5N	464.25	260.45	4.1796	21.7958	Himmele, W. et al (1979)
N,N-dimethylaniline $C_8H_{11}N$	467.25	275.55	1.6798	20.0755	Himmele, W. et al (1979)
Water	373.15	273.15	1.8495	47.8127	Ginnasi , A. et al (1976)

Source of thermophysical properties is ICAS Software version 13 (2010)

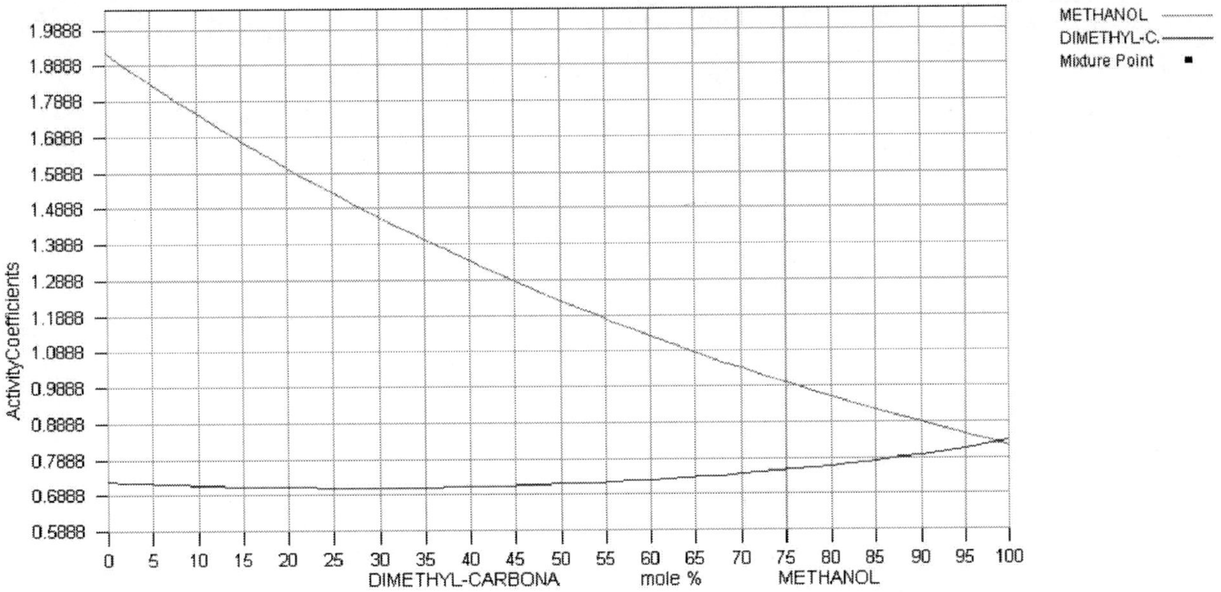

Figure 1: Activity coeff. at 1 atm , 370K , 50% mol phenol with azeotrope feed 87.06 mol MeOH;12.94 mol DMC

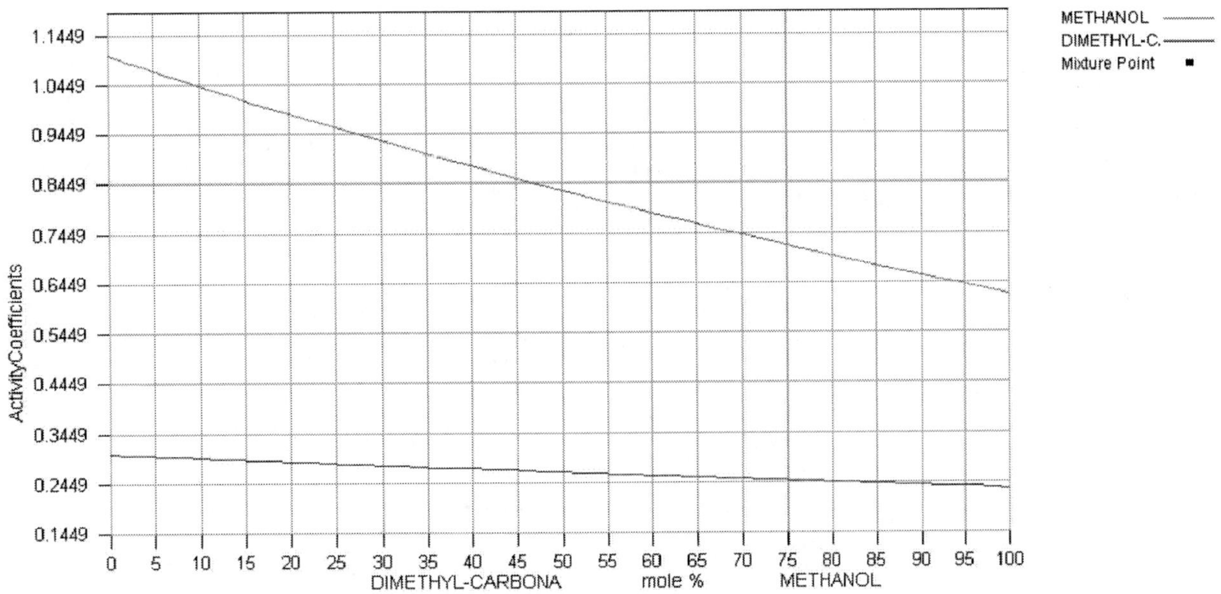

Figure 2: Activity coeff. at 1 atm , 370K , 80% mol phenol with azeotrope feed 87.06 mol MeOH;12.94 mol DMC

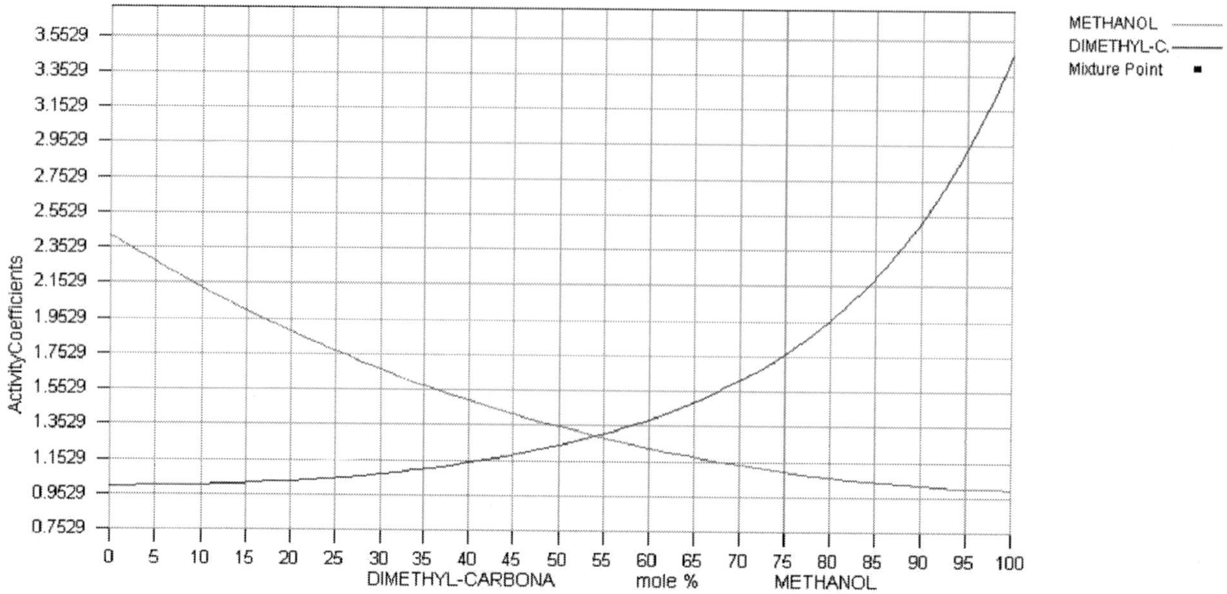

Figure 3: Activity coeff. at 1 atm , 370K , 0 % mol aniline with azeotrope feed 87.06 mol MeOH;12.94 mol DMC

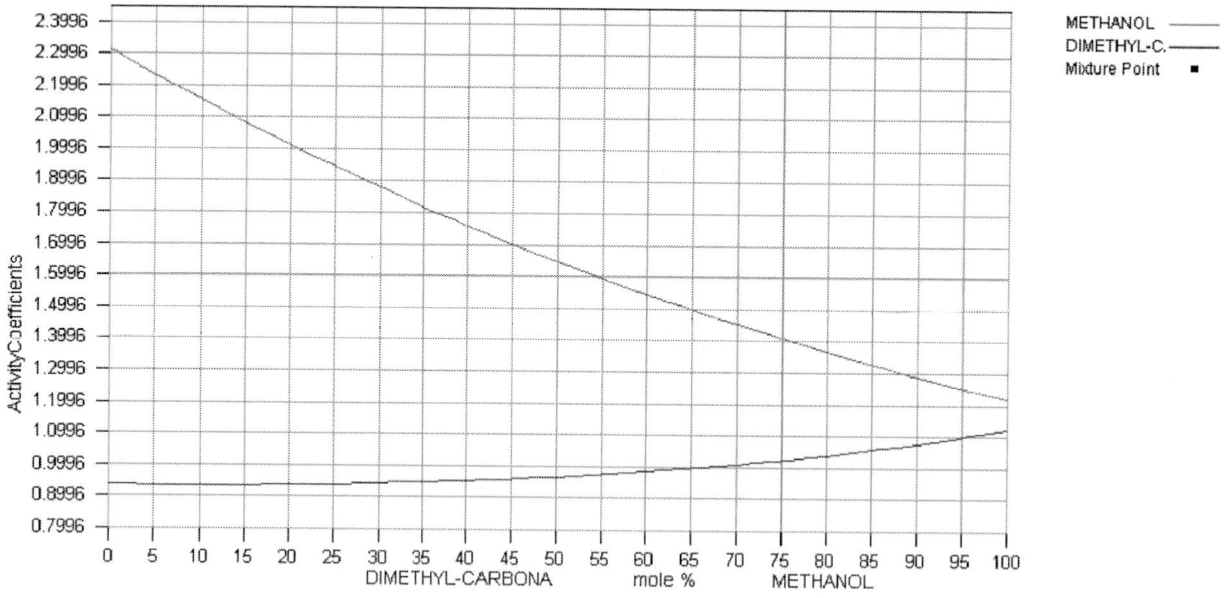

Figure 4: Activity coeff. at 1 atm , 370K , 50% mol aniline with azeotrope feed 87.06 mol MeOH;12.94 mol DMC

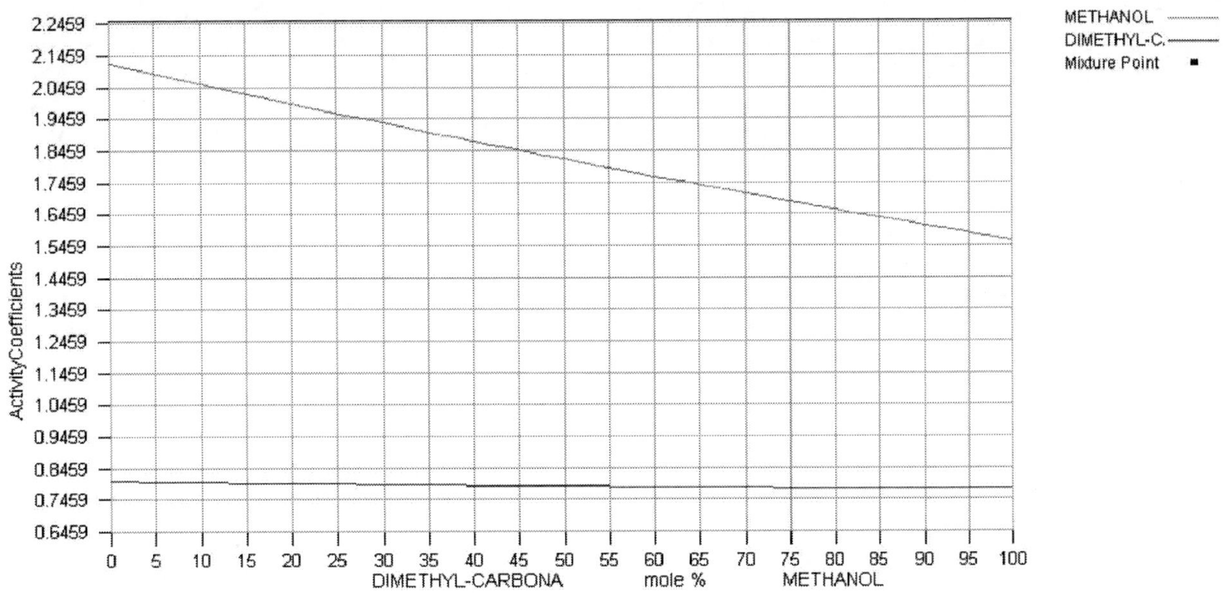

Figure 5: Activity coeff. at 1 atm , 370K , 80% mol aniline with azeotrope feed 87.06 mol MeOH;12.94 mol DMC

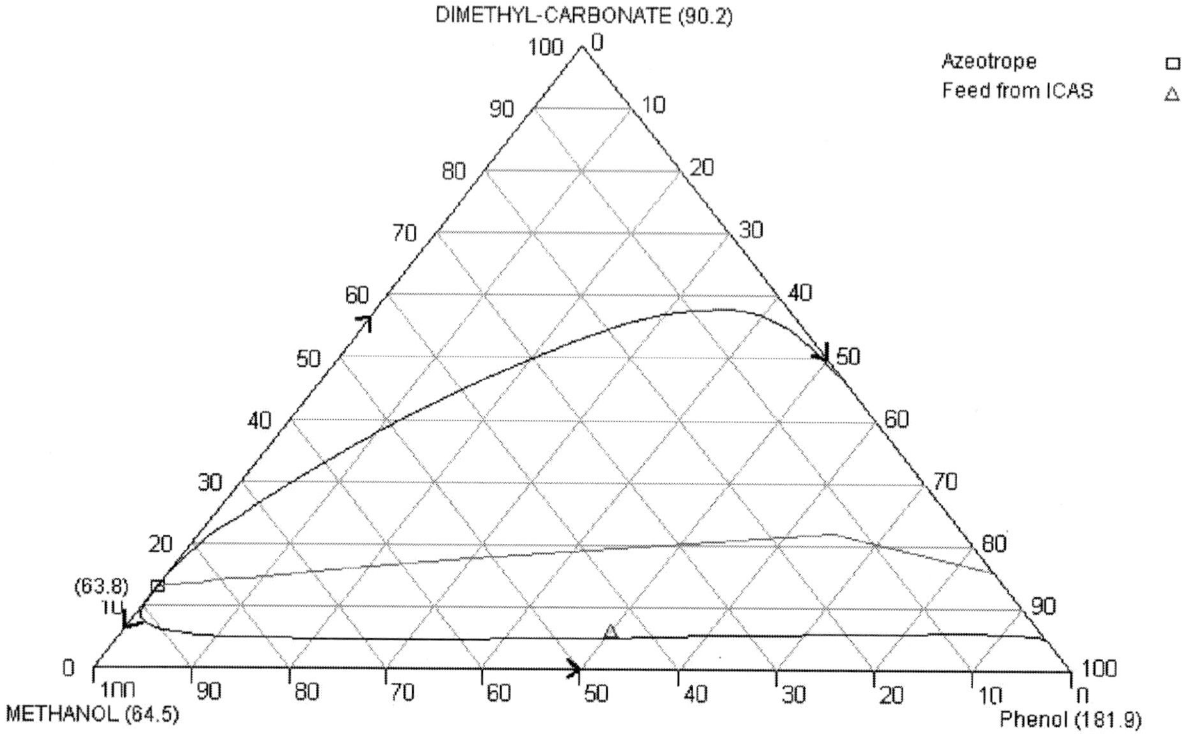

Upper Binodal Curve at 1= 30%, 2=50%, and 3= 20%

Lower Binodal Curve at 1=50%, 2=5%, 3= 45%

Distillation Boundary

Feed from ICAS: 1=87.06 mole, 2= 12.94 mole , and 3 = 100 mole (Azeotrope feed MeOH-DMC)

1=mol % Methanol, 2 = mole % DMC , 3 = mol % Phenol

Figure 6: Residue Curve Map for Extractive Distillation of Methanol-DMC-Phenol System at 1 atm

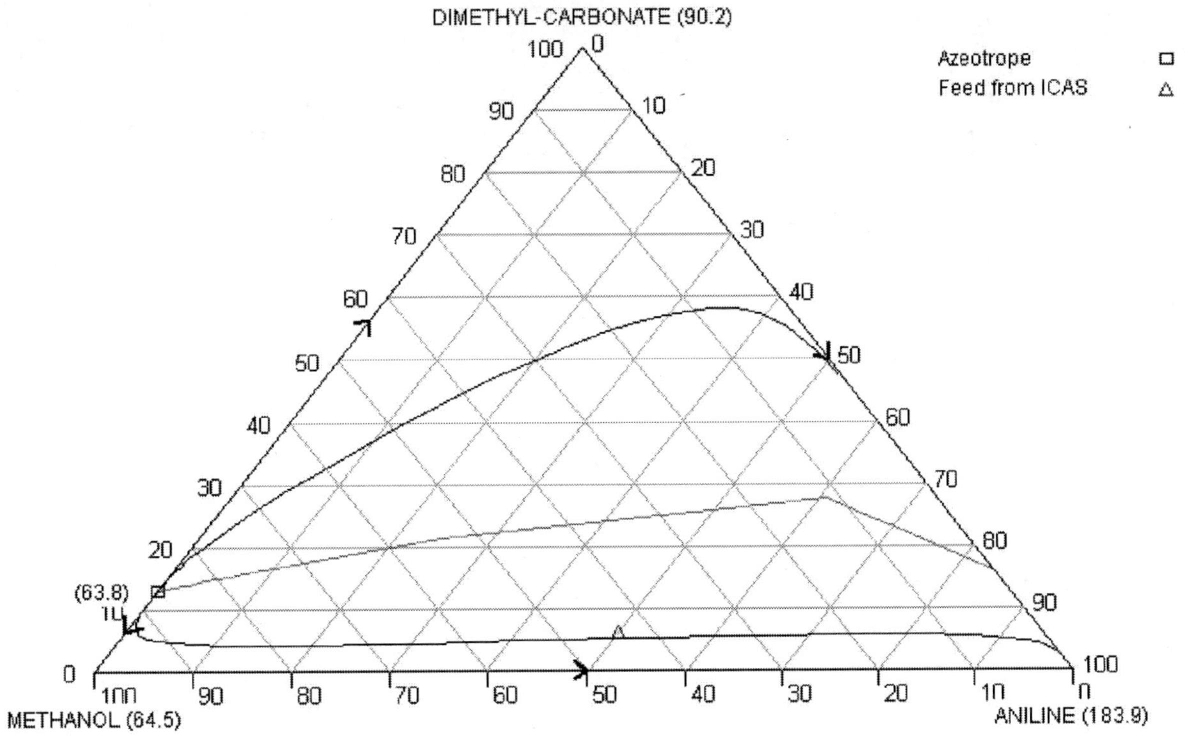

______________ Upper Binodal Curve at 1= 30%, 2=50%, and 3= 20%

______________ Lower Binodal Curve at 1=50%, 2=5%, 3= 45%

______________ Distillation Boundary

Feed from ICAS: 1=87.06 mole, 2= 12.94 mole , and 3 = 100 mole (Azeotrope feed MeOH-DMC)

1=mol % Methanol, 2 = mole % DMC , 3 = mol % Aniline

Figure 7: Residue Curve Map for Extractive Distillation of Methanol-DMC-Aniline System at 1 atm

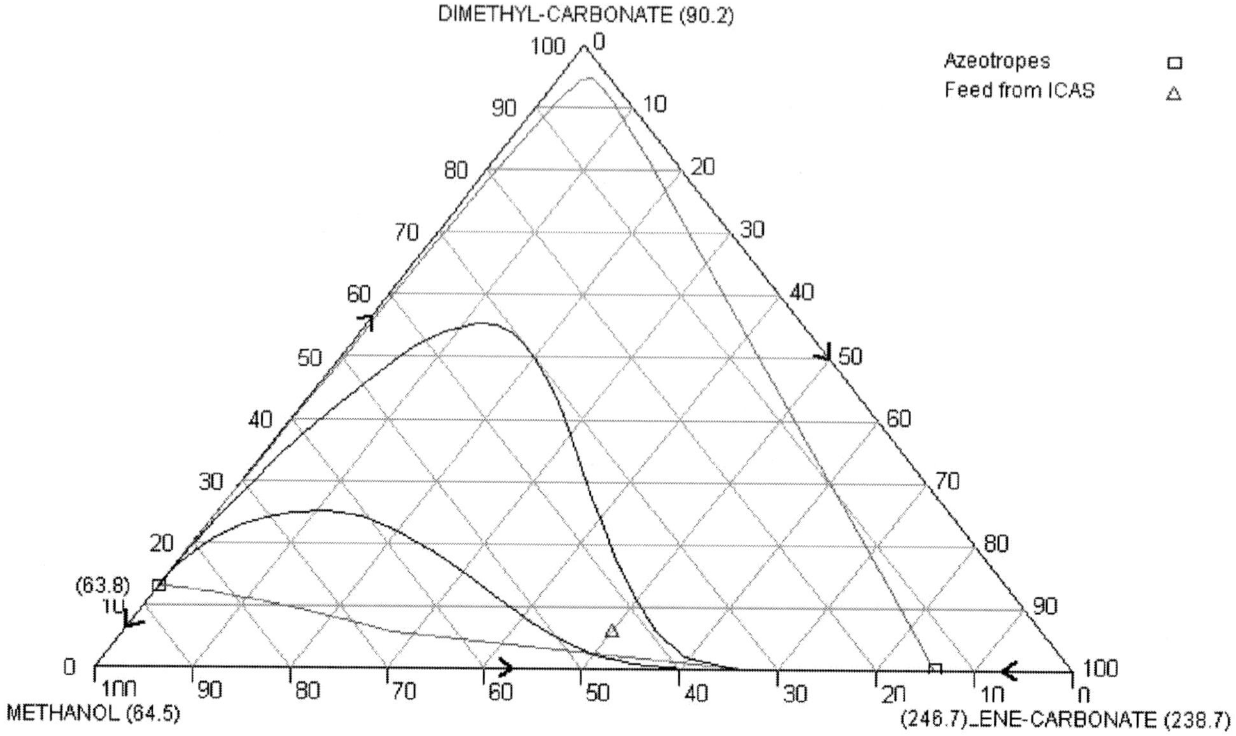

______________ Upper Binodal Curve at 1= 30%, 2=50%, and 3= 20%

______________ Lower Binodal Curve at 1=50%, 2=5%, 3= 45%

______________Distillation Boundary

Feed from ICAS: 1=87.06 mole, 2= 12.94 mole , and 3 = 100 mole (Azeotrope feed MeOH-DMC)

1=mol % Methanol, 2 = mole % DMC , 3 = mol % Ethylene Carbonate

Figure 8: Residue Curve Map for Extractive Distillation of Methanol-DMC-Ethylene Carbonate System at 1 atm.

Conclusion:

The focus of this project is the application of a heat-integrated reactive distillation couple with a heat-integrated extractive distillation process for the production of a high purity dimethyl carbonate by transesterification of ethylene carbonate with methanol to achieve energy reduction/savings and is economically promising.

References:

Fang, Y. J.; Qian, J. M. Isobaric vapor-liquid equlibria of binary mixtures containing the carbonate group -OCOO-. *J. Chem. Eng. Data* **2005**, *50*, 340–343.

Fang, Y.; Xiao, W.; Experimental and modeling studies on a homogeneous reactive distillation system for dimethyl carbonate synthesis by transesterification. Separation and Purification Technology 34 (2004) 255-263
Fang, Y. J.; Xiao, W. D. Experimental and modeling studies on a homogeneous reactive distillation system for dimethyl carbonate synthesis by transesterification. *Sep. Purif. Technol.* **2004**, *34*, 255–263.

Foucher, E. R.; Doherty, M. F.; Malone, M. F. Automatic screening of entrainers for homogeneous azeotropic distillation. *Ind. Eng. Chem. Res.* **1991**, *30*, 760–772.

Gmehling, J.; Mollmann, C. Synthesis of distillation processes using thermodynamic models and the Dortmund data bank. *Ind. Eng. Chem. Res.* **1998**, *37*, 3112–3123.

Ginnasi, A., Passoni, G.; Extractive distillation of a dimethyl carbonate feed with water. Patent US 3,963,586 (196)

Harmsen, G. Jan, Reactive Distillation: The front-runner of industrial process intensification. A full review of commercial applications, research, scale-up, design and operation. Chemical Engineering and Processing 46(2007) 774-780

Himmele, W. et al; Preparation of pure dimethyl carbonate, patent US 4,162,200 (1979)

Hiwale, R.S.; Bhate, N. V.; Mahajan, Y.S.; Mahajani, S. M.; Industrial applications of reactive distillation:Recent trends. International Journal of Chemmical Reactor Engineering, Vol 2(2004), Review R1 pp 1- 52

Hsu, K.; Hsiao, Y.; Chien, I.; Design and control of dimethyl carbonate-methanol separation via extractive distillation in dimethyl carbonate reactive-distillation process. Ind. Eng. Chem. Res. 210, 49, 735-749

http://www.chemsystems.com/reports/search/docs/abstracts/99S6-abs.pdf ; Developments in Dimethyl Carbonate Production Technologies PERP-99-S6 (May. 2000)

Hu, Wang-Hu; Shen, L; Zhao, L, Measurement of vapor-liquid equilibrium of binary mixtures of phenol-dimethyl carbonate and phenol-methanol at 101.3 kPa. Fluid Phase Equilibria 219 (2004) 265-268

ICAS Software version 13, CAPEC, Technical University of Denmark (2010)

Kossack, S.; Kramer, K.; Gani, R.; Marquardt, W. A systematic synthesis framework for extractive distillation processes. *Chem. Eng. Res. Des.* **2008**, *86*, 781–792.

Lek-utaiwan, P; Suphanit, B; Mongkolsiri, N.; Gani, R.; Integrated design of solvent-based extractive separation processes including experimental validation. 19th European Symposium on Computer Aided Process Engineering-ESCCAPE 19 J.Jezowski and J. Thullie (Editors), 2009 Elsevier

Li, C.; Zhang, S. Environmental Benign design of DMC production process. Trans IChemE, Part A, Chemical Engineering Research and Design, 2006, 84(A1):1-8

Nishihira, K., Yoshida, S; Tanaka, S., Continuous process for preparing dimethyl carbonate. Patent US 5,124,185 (1993)

Pasternak, M., et al; Membrane separation of organic liquids, patent US 4,960,519 (1990).

Rodriguez, A.; Canosa, J.; Dominguez, A.; Tojo, J. Vapour-liquid equilibria of dimethyl carbonate with linear alcohols and estimation of interaction parameters for the UNIFAC and ASOG method. *Fluid Phase Equilib.* **2002**, *201*, 187–201.

Rodriguez-Donis, I.; Gerbaud, V.; Joulia, X. Entrainer selection rules for the separation of azeotropic and close-boiling-temperature mixtures by homogeneous batch distillation process. *Ind. Eng. Chem. Res.* **2001**, *40*, 2729–2741.

Ryu, J. V.; Gelbein, A. P.; Process and catalyst for making dialkyl carbonates, patent US 6,392,078 B1 (2002)

Sachs, H. M.; Preparation of alkylene carbonates. Patent US 4,786,741 (1988)

Wang, S. J.; Yu, C. C.; Huang, H. P. Plant-wide design and control of DMC synthesis process via reactive distillation and thermally coupled extractive distillation *Comput. Chem. Eng.* 34(2010) 361-373

Wang, X; Yang, B.; Wang, D.; Zhai X.; Reactive rectifying for producing dimethyl carbonate. Chemical Engineering Journal 122 (2006) 15-20

(77b) Reactive Distillation Process for Glycerol Ethers Production

Elena Vlad[1], **Costin Sorin Bildea**[2], Marina Mihalachi[1] and Grigore Bozga[1], (1)Department of Chemical Engineering, University Politehnica of Bucharest, Bucharest, Romania, (2)Department of Chemical Engineering, University POLITEHNICA Bucharest, Bucharest, Romania

Glycerol is produced as by-product of biodiesel production in amounts that are equivalent to approximately 10 % wt. of the total product. As a result to the increasing availability, the market price of glycerol has dropped rapidly and new uses for glycerol need to be found. Di- and tri-ethers of glycerol are compounds soluble in diesel and biodiesel, improving the quality of the fuel. They diminish the emissions of particulate matter, carbon oxide and carbonyl compounds and decrease the cloud point of diesel fuel when combined with biodiesel. Therefore, ethers of glycerol are interesting alternatives to commercial oxygenate additives.

In this contribution, we demonstrate the feasibility of a reactive distillation process for production of glycerol ethers by etherification of glycerol with iso-butene catalyzed by heterogeneous acid catalysts, such as Amberlist 35. The process is performed in a reactive distillation column, with few un-reactive trays placed above and below the reactive zone. Glycerol is fed as liquid at the top of the reactive zone. Iso-butene enters in the column as a vapor stream at the bottom. The bottom product contains the di- and tri-ethers of glycerol which are the main products. The vapor outlet stream contains the mono-ether and the excess of iso-butene, which are separated in a decanter. The mono-ether is returned as reflux, while the excess of iso-butene is recycled. The detailed design is performed using AspenPlus. The stability and robustness in operation is checked using rigorous dynamic simulation in AspenDynamics.

See more of this Session: Process Intensification (II)
See more of this Group/Topical: Process Development Division

(77c) Advanced Reactive Distillation Concepts for the Indirect Hydration of Cyclohexene to Cyclohexanol

Aspi K. Kolah[1], Lars Peereboom[1], Carl T. Lira[1], Jing Huang[2], C.B. Panchal[2], Robert W. Lyczkowski[2], Emmanuel A. Dada[3], Richard D. Doctor[2] and Dennis J. Miller[1], (1)Chemical Engineering and Materials Science, Michigan State University, East Lansing, MI, (2)Energy System Division, Argonne National Laboratory, Argonne, IL, (3)ChemProcess Technologies (CPT), LLC, League City, TX

Cyclohexanol is a very important bulk chemical feedstock for the polymer industry, primarily as a precursor to nylon but also a plasticizer component and solvent. The conventional process for cyclohexanol synthesis is either cyclohexane oxidation, hydrogenation of phenol, or the direct hydration of cyclohexene. The direct hydration route, which is the simplest in principle, suffers from limited solubilities of the cyclohexene-water pair, thus requiring large amounts of catalyst and high residence times.

An alternative to the direct hydration of cyclohexene is the indirect hydration process where acetic acid is used as the reactive entrainer. In the first step of this process cyclohexene from the inlet mixture of cyclohexene and cyclohexane reacts with acetic acid to form cyclohexyl acetate, which is hydrolyzed with water to cyclohexanol in the second step.

In this paper we present the reaction of cyclohexene from a mixture of cyclohexene and cyclohexane with acetic acid to form cyclohexyl acetate in the MSU pilot scale reactive distillation column packed with Katapak-SP11 structured packing. Effect of flow rates, reflux ratio and column geometry have been experimentally investigated. Column performance is predicted using RADFRAC in Aspen Plus process simulation software with appropriate kinetic parameters derived from experimental batch kinetic studies. Additionally, single- and two-phase flow through a Katapak-SP11 structured packing was modeled using ANSYS FLUENT 12.1. The purpose of this analysis is to determine the liquid flow distributions interior and exterior to the catalyst baskets, catalyst wetting, and hence reaction effectiveness. Advanced concepts such as heat integration are incorporated into experimental and simulation results. Plant scale simulations are presented.

See more of this Session: Process Intensification (II)
See more of this Group/Topical: Process Development Division

A review of techniques for the process integration of coupling exothermic and endothermic reactions

Mohammad Reza Rahimpour;
Davood Iranshahi;
majid raoof Dehnavy;
Fatemeh Allahgholipour;

Presentation ID:212277

School of Chemical and Petroleum Engineering, Shiraz University, P.O. Box 71345, Shiraz, Iran

Keywords: Couple, Endothermic, Exothermic, Process integration

Corresponding author. Tel.: +98 711 2303071; fax: +98 711 6287294.
E-mail address: rahimpor@shirazu.ac.ir (M.R. Rahimpour).

1. Introduction

Reducing environmental pollutions, reducing capital and operating costs of processes and increasing efficiency and outputs of processes are significant terms which should be noticed in this new world. Process integration (PI), which is one of the most significant methodology in chemical engineering and process technology, can prove these aims by combining several processes to reduce consumption of the energy and material or environmental emissions [1] and also it can decrease the size of equipment by several orders of magnitude leading to substantial savings in capital cost or improvement of intrinsic safety. The study on process integration started in the late 1970s and early 1980s. [2]

PI includes hierarchical design methods, knowledge-based systems, numerical and graphical techniques and pinch analysis [3]. Pinch analysis designs a process to minimize energy consumption and enhance energy efficiency. In this way multifunctional reactor is a new one [4]. The best field of using multifunctional reactors is the coupling of endothermic and exothermic reactions. In this way, an exothermic reaction(s) produces the source of heat to drive the endothermic reaction(s). The distinction between net source and net sink regions is a key point in the pinch analysis [3]. In this paper, we are investigating review of coupling concept with focus on the reactors, and also, consequences of previous researches in this field are evaluated.

2. Categorize the kinds of coupling of exothermic and endothermic [5]:

2. a- Direct coupling (direct coupled adiabatic reactor)
2. b- Regenerative coupling (reverse-flow reactor)
2. c- Recuperative coupling (counter current heat exchanger reactor and co-current heat exchanger reactor).
2. d- Membrane reactor

2. a) In directly coupled reactors, exothermic and endothermic reactions are in a same bed. There is a direct heat transfer between them, and also, both reactions occur simultaneously. Direct coupling adiabatic reactor (DCAR) can be also classified into simultaneous DCAR (SIMDCAR) and sequential DCAR (SEQDCAR). In SIMDCAR, uniformly mixed endothermic and exothermic catalysts make the bed of it. As it can be seen in Fig.1 (a) cayalyst are in a same bed and feed contains both of reactants (exothermic reactant and endothermic one). Fig.1 (b) illustrates a SEQDCAR reactor.

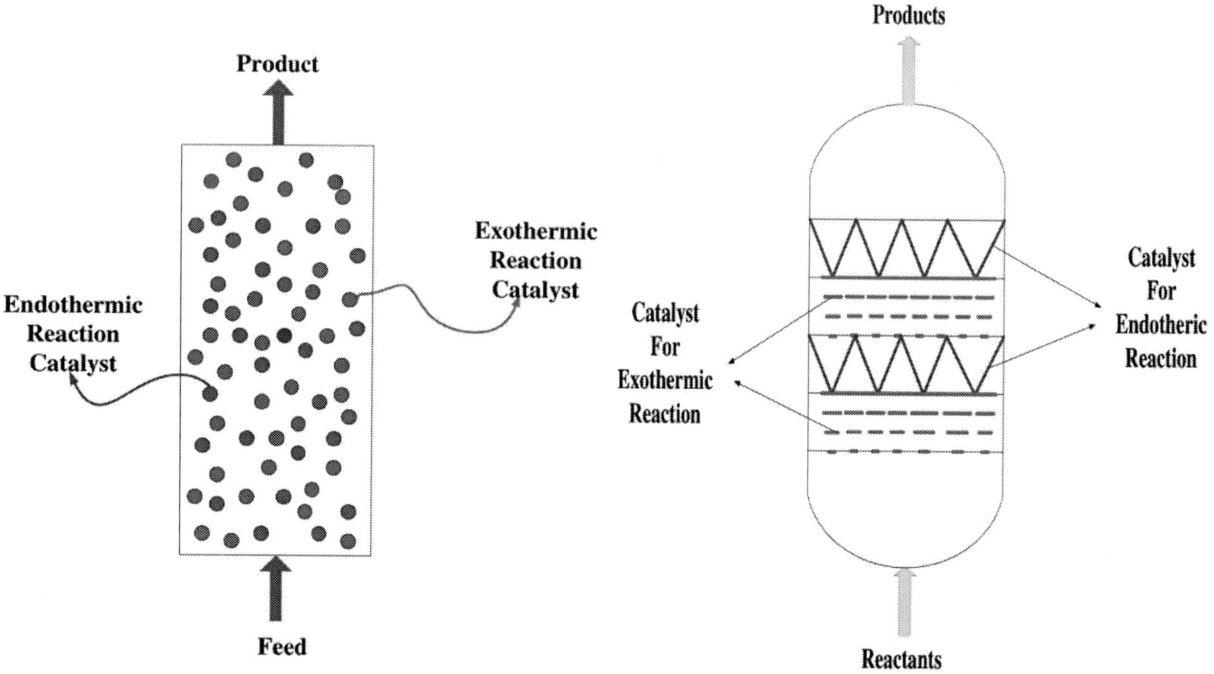

Figure.1 (a) Simultaneous direct coupled adiabatic reactor and (b) sequential direct coupled adiabatic reactor.

2. b) In regenerative coupling reactor, both reactions are in a same place but separated in time. There are two steps and exothermic (or endothermic) and endothermic (or exothermic) reactants enter the reactor respectively. Regenerative coupling reactors are well adaptable for weakly exothermic reactions such as the catalytic purification of exhaustive streams are polluted by volatile organic compounds (Kulkurani, 1996). In the first half of the cycle, the exothermic reaction takes place. Then, the flow is reversed in the next half of the cycle and the endothermic reaction occurs by using of the stored energy in the bed during the previous exothermic half cycle.

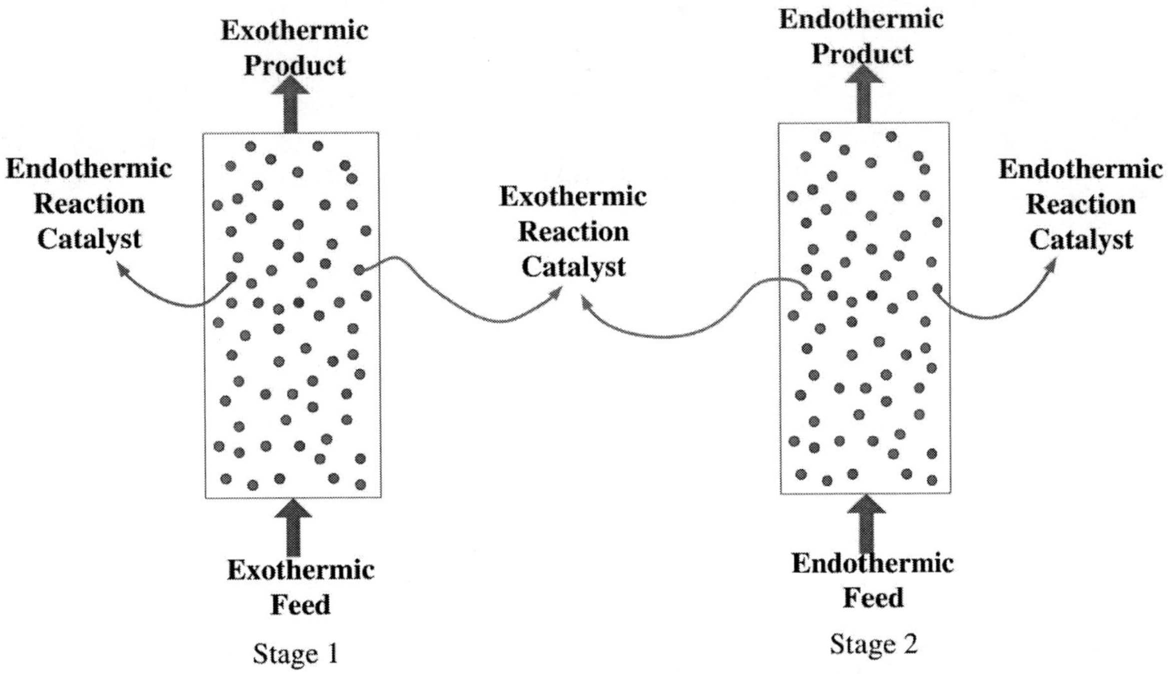

Figure.2 Regenerative coupling reactor.

2. c) In recuperative coupling reactor that is showed in fig.3, there are two separated spaces for two reactions. Tube side is for endothermic reaction and shell side is for exothermic one and heat is transferred indirectly between shell and tube side. In this kind of reactor, the reactors for the endothermic reactions are fed through contiguous passages of the reactor [6, 7]. These passages are either channels in a monolith or catalyst coated. Koga and Watanabe (1991) described a coupling of the catalytic combustion and reforming, which took place in alternate channels. Igurshi et al., (1992) described a wall reactor that supplied heat to endothermic reaction with a heating medium.

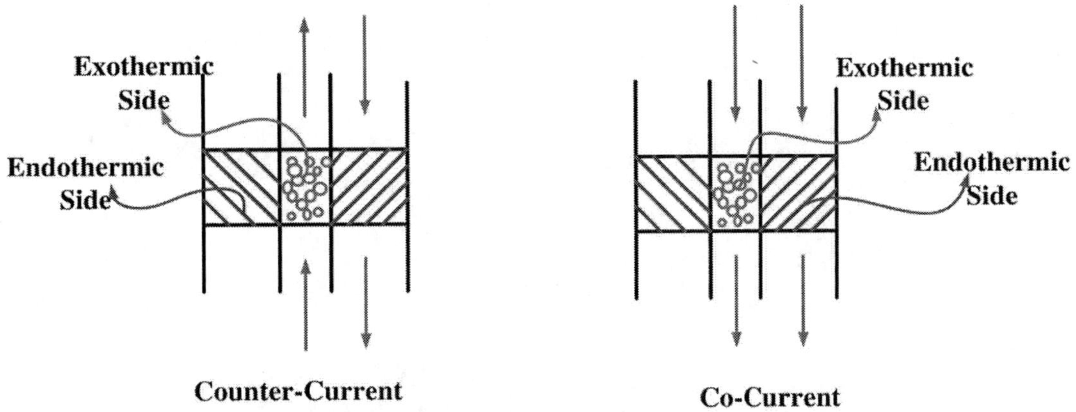

Figure.3 Recuperative coupling reactor.

2. d) In membrane reactor, exothermic reaction carries out in the tube side, and endothermic reaction occurs in through shell side and membrane. A simplified schematic of the membrane reactor is illustrated in Fig.4.

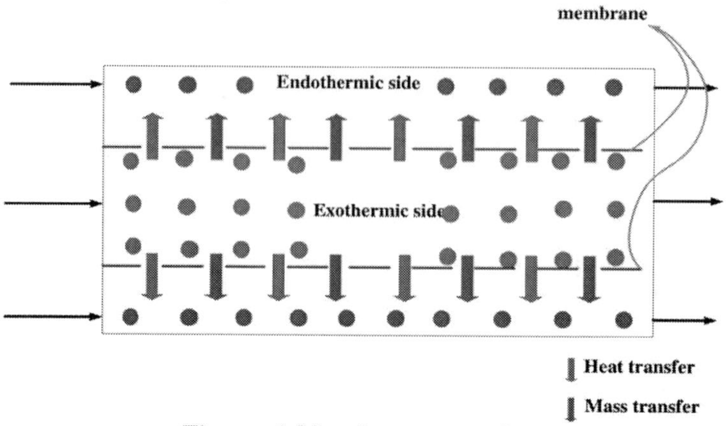

Figure.4 Membrane reactor.

3. Main Table

In this section, we demonstrate examples of the various studies on coupling concept in the Main Table.

No.	Sec.	Reaction	Reaction kinetic	ΔH_{298} (Kj/mol)	T (K)	P	Phase	Catalyst	Author	Ref
1	1-1	Methane combustion reaction	$CH_4 + 2O_2 \rightarrow CO_2 + 2H_2O$	(-)	1000	3 bar	G	Pt / Al_2O_3	Annaland et al.	9, 10, 11, 12
	1-2	Dehydrogenation Propane to propylene	$C_3H_8 \leftrightarrow C_3H_6 + H_2$	130	823-923	1 atm	G			
2	2-1	Methanol Synthesis	$CO + H_2 \rightarrow CH_3OH$	(-)	500	76.98 bar	G	CuO / ZnO / Al_2O_3	Rahimpour et al.	6, 13, 14, 15, 23
	2-2	Dehydrogenation of Cyclohexane to benzene	$C_6H_{12} \rightarrow C_3H_6 + 3H_2$	+206.2.	423-523	101.3 KPa	G	Pt / Al_2O_3		

3	**3-1**	Dehydrogenation methanol to (DME)	$2CH_3OH \rightarrow CH_3OCH_3 + H_2O$	-23.4	535	18.18 bar	G	γ / Al_2O_3	Rahimpour et al.	16
	3-2	Dehydrogenation of Cyclohexane to benzene	$C_6H_{12} \rightarrow C_3H_6 + 3H_2$	206.2	423-523	101.3 KPa	G	Pt / Al_2O_3		
4	**4-1**	Hydrogenation of nitrobenzene to aniline	$C_6H_5NO_2 + 3H_2 \rightarrow C_6H_5NH_2 + 2H_{2O}$	-443	780	1100 KPa	G	Pd / Al_2O_3	Iranshahi et al.	8
	4-2	Naphtha reforming	(CnH2n) ↔ (CnH2n-6) + 3H2	71038.06	777	3703 KPa	G	Pt / Re / Al_2O_3		
			(CnH2n)+H2 ↔ (CnH2n+2)	-36395.33						
			(CnH2n) + n/3H2 → (C1–C5)	-51939.31						
			(CnH2n+2) + (n−3)/3H2 → (C1–C5)	-56597.54						
5	**5-1**	Hydrogenation of nitrobenzene to aniline	$C_6H_5NO_2 + 3H_2 \rightarrow C_6H_5NH_2 + 2H_{2O}$	−443	300 - 700	1.1 bar	G	Pd / Al_2O_3	Abo-Ghandar et al. & Qin et al.	17, 18, 19
	5-2	Dehydrogenation of ethylbenzene to styrene	$C_6H_5CH_2 \leftrightarrow C_6H_5CHCH_2 + H_2$	+117.6	300 - 700	4.5 bar	G	Fe_2O_3		

6	**6-1**	The CO2 methanation	$CO_2 + 4H_2 \leftrightarrow CH_4 + 2H_2O$	-165	300 – 700	0.1-1 MPa	G	Ni	Qin et al.	19
	6-2	Dehydrogenation of ethylbenzene to styrene	$C_6H_5CH_2 \leftrightarrow C_6H_5CHCH_2 + H_2$	+117.6	300 - 700	4.5 bar	G	iron oxide		
7	**7-1**	Methane oxidation	$CH_4 + 2O_2 \rightarrow CO_2 + 2H_2O$	-802	300 – 700	15 – 25 bar	G	Ni	Abashar & Elnashei	20, 21, 22
	7-2	Methane steam reforming	$CH_4 + H_2O \rightarrow CO + 3H_2$	+206			G			

The most significant authors related with coupling of exothermic and endothermic reactions field are Annaland, Abo-Ghander, Rahimpour, Abashar, Elnashaie. Work of these head authors were extended or used by other researchers.
By investigation on the authors work which are explained in the Main Table, It can be seen most of them have utilized membrane and recuperative model in their work

4. Conclusion

With this literature review we aimed to give information about coupling the exothermic and endothermic reactions to researchers. The papers that were listed in the Main Table revealed values of the coupling of exothermic and endothermic reactions and coupled reactors. Nowadays, process integration, particularly pinch analysis is the inseparable part in the chemical processes. Therefore, the coupling techniques as a pinch analysis should be taken into consideration and used in chemical industries. As it can be seen in all work related with this field, the short distance between heat sink and heat source could increase efficiency of the heat transfer and incorporating two separated processes with a coupled reactors could reduce size of the operations. And also there are still many reactions which can be coupled with each other. Studying on new coupling of reactions is needed in order to achieve PI's aims. This shows priority of the novel models (coupled reactors configuration) to the conventional configurations.

References

1. F. Friedler (2010), Process integration, modeling and optimization for energy saving and pollution reduction, Applied Thermal Engineering (2008), doi: 10.1016/ j.applthermaleng. 04.030
2. R. Smith(2000), State of the art in process integration, Applied Thermal Engineering (2008), 20 1337-1345
3. A.P. Rossiter(2003), Succeeding in process integration, Process Industries Expo User Conference, March 31.
4. L. Lavric, V. Plesu, J. De Ruyck(2005), Chemical reactors energy integration through virtual heat exchangers-benefits and drawbacks, Applied Thermal Engineering1033-1044.
5. Ramaswamy C. Ramaswamy(2006), Steady state and dynamic reactor models for coupling exothermic and endothermic reactions, doctor's degree at the University of Washington, May,.
6. M.H. Khademi, M.R. Rahimpour, A. Jahanmiri(2010), Differential evolution (DE) strategy for optimization of hydrogen production, cyclohexane dehydrogenation and methanol synthesis in a hydrogen-permselective membrane thermally coupled reactor, Int.J.Hydrogen Energy 35 1936 – 1950
7. G. Kolios, J. Frauhammer, G. Eigenberger(2001), A simplified procedure for the optimal design of autothermal reactors for endothermic high-temperature reactions., Chem. Eng. Sci. 56 351-357
8. D. Iranshahi, A.M. Bahmanpour, E. Pourazadi, M.R. Rahimpour(2010), Mathematical modeling of a multi-stage naphtha reforming process using novel thermally coupled recuperative reactors to enhance aromatic production, Int.J.Hydrogen Energy 35 10984-10993
9. M.van Sint Annaland, J.A.M. Kuipers, W.P.M.van Swaaij56 (2001), Safety analysis of switching between reductive and oxidative conditions in a reaction coupling reverse flow reactor. Chem.Eng.Sci. 1517-1524
10. M.van Sint Annaland, H.A.R. Scholts, J.A.M. Kuipers, W.P.M.van Swaaij57 (2002), A novel reverse flow reactor coupling endothermic and exothermic reactions. Part I: comparison of reactor configurations for irreversible endothermic reactions, Chem.Eng.Sci. 833 – 854
11. M.van Sint Annaland, H.A.R. Scholts, J.A.M. Kuipers, W.P.M.van Swaaij57 (2002), A novel reverse flow reactor coupling endothermic and exothermic reactions. Part II: Sequential reactor configuration for reversible endothermic reactions, Chem.Eng.Sci. 855 – 872
12. M.van Sint Annaland, R.C. Nijssen57 (2002), A novel reverse flow reactor coupling endothermic and exothermic reactions: an experimental study, Chem.Eng.Sci. 4967 – 4985
13. M.H. Khademi, A. Jahanmiri, M.R. Rahimpour34 (2009), A novel configuration for hydrogen production from coupling of methanol and benzene synthesis in a hydrogen-permselective membrane reactor, Int.J.Hydrogen Energy 5091 – 5107
14. M.H. Khademi, P. Setoodeh, M.R. Rahimpour, A. Jahanmiri34 (2009), Optimization of methanol synthesis and cyclohexane dehydrogenation in a thermally coupled reactor using differential evolution (DE) method, Int.J.Hydrogen Energy 6930 – 6944
15. M.R. Rahimpour, E. Pourazadi(2010), A Comparison of hydrogen and methanol production in a thermally coupled membrane reactor for co current and counter current flows, Int.J.Energy Research DOI: 10.1002/er.1744
16. M. Farsi, M.H. Khademi, A. Jahanmiri, M.R. Rahimpour(2010), Novel Recuperative Configuration for Coupling of Methanol Dehydration to Dimethyl Ether with Cyclohexane Dehydrogenation to Benzene, Ind. Eng. Chem. Res. 49, 4633–4643

17. N. S.Abo-Ghander, F. Logist, J.R. Grace, J.F.M. V. Impe , S.S.E.H. Elnashaie, C.J. Lim(2010)
, Optimal design of an autothermal membrane reactor coupling the dehydrogenation of ethylbenzene to styrene with the hydrogenation of nitrobenzene to aniline, Chem.Eng.Sci
18. N.S. Abo-Ghandera, J.R. Gracea, S.S.E.H. Elnashaie, C.J. Lim63 (2008), Modeling of a novel membrane reactor to integrate dehydrogenation of ethylbenzene to styrene with hydrogenation of nitrobenzene to aniline, Chem.Eng.Sci, 1817 – 1826
19. Z. Qin, J. Liu, A. Sun, J. Wang(2003), Reaction Coupling in the New Processes for Producing Styrene from Ethylbenzene, Ind. Eng. Chem. Res. 42, 1329-1333
20. M.E.E. Abashar, S.S.E.H. Elnashaie(2007), Feeding of oxygen along the height of a circulation fast fluidized bed membrane reactor for efficient production of hydrogen, Trans IChemE, Part A, Chemical Engineering Research and Design, 85(A11): 1529–1538
21. Y. Chen, H. Xu, X. Jin, G. Xiong116 (2006), Integration of gasoline prereforming into autothermal reforming for hydrogen production, Catalysis Today 334–340
22. M.E.E. Abashar, F.M. Alhabdana, S.S.E.H. Elnashaie33 (2008), Discrete injection of oxygen enhances hydrogen production in circulating fast fluidized bed membrane reactors, Int.J.Hydrogen Energy 2477 – 2488
23. M.R. Rahimpour, F. Rahmani, M. Bayat, E. Pourazad(2010), Enhancement of simultaneous hydrogen productionand methanol synthesis in thermally coupleddouble-membrane reactor, Int.J.Hydrogen Energy xxx 1 -1 5

2011

(88a) Catalytic Formation of Carbonyl Sulfide During Treatment of Coal-Derived Fuel Gas with Pd/γ-Al_2O_3 Sorbents

Erik C. Rupp, Evan J. Granite and Dennis C. Stanko, National Energy Technology Laboratory, Pittsburgh, PA

This group has developed a Pd/γ-Al_2O_3 sorbent capable of removing trace contaminants at moderate to high temperatures (200 – 450 °C) from coal-derived fuel gas. In the development of the sorbent, the catalytic formation of carbonyl sulfide (COS) was noted, which would require modification to sulfur removal processes. The formation of COS over a Pd catalyst is also a major concern for Pd alloy based hydrogen separation membranes proposed for use with fuel gas. The conditions required for catalytic formation have been explored, with up to 80 ppmv COS formed under gas-solid contact times of under 0.1 s. A reaction mechanism is proposed and multiple kinetic models are explored based on experimental results. The potential for hydrolysis of COS based on the γ-Al_2O_3 support is considered.

See more of this Session: Clean Coal Technology
See more of this Group/Topical: Energy and Transport Processes

(88b) Sustainability Assessment of Polygeneration Systems

Abhishek Jayswal[1], Dr. Helen Lou[2], Preeti Gangadharan[1], Kailang Zeng[1] and Anand Zambar[1], (1)Dan F. Smith Department of Chemical Engineering, Lamar University, Beaumont, TX, (2)Chemical Engineering, Lamar University, Beaumont, TX

Production of chemicals, electricity and fuels from natural resources via polygeneration systems is one of the emerging clean technologies for reducing greenhouse emissions. The polygeneration systems unite chemical, energy and fuel sectors and provide excellent efficiencies, in terms of materials and energy conversion by receiving multiple natural resources as feeds and producing many useful products through various interrelated processes. The complex nature of polygeneration systems presents a challenge for analysis of their sustainability footprints. This study presents a systematic approach for sustainability assessment of such processes by merging exergy efficiency analysis and inherently safe design approach into conventional economic, environmental and social aspects of sustainability. This work compares several different Polygeneration systems, which uses coal and natural gas as feeds, on sustainability criteria.

See more of this Session: Clean Coal Technology
See more of this Group/Topical: Energy and Transport Processes

(88c) Analysis of Freshwater Usage In Advanced Gasification Based Power Systems

Michael E. Walker and Javad Abbasian, Chemical and Biological Engineering, Illinois Institute of Technology, Chicago, IL

The thermoelectric power industry in the U.S. uses a large amount of freshwater. The large water demand is increasingly a problem, especially for new power plant development, as the availability of freshwater diminishes in the United States. Gasification based systems are next generation coal conversion processes that achieve higher process efficiencies than conventional power systems. Furthermore, gasification based systems use significantly less water than PC-Boiler systems, particularly when carbon dioxide capture strategies are applied.

A wide array of work has been focused on reducing water usage in existing PC plants. This paper addresses the applicability of developed water conservation methods to gasification based systems, e.g. Integrated Gasification Combined Cycle (IGCC). This work will provide insight into the challenges and limitations of water reduction within the IGCC system through careful analysis of detailed process models. In addition, water savings from novel process configurations will be discussed.

See more of this Session: Clean Coal Technology
See more of this Group/Topical: Energy and Transport Processes

DECAB: Process development of a phase change absorption process

E. Sanchez Fernandez, E. Kimball, E.L.V. Goetheer
TNO Separation Technology
Delft, the Netherlands

Introduction

This work describes a novel separation process to remove CO_2 from flue gas based on precipitating solvents (DECAB). The reaction of CO_2 with alkanol amines has been extensively described in the literature ([1], [2], [3], [4]). Depending on the type of amine, different species are predominantly formed. In the case of primary amines, the favored species is carbamate, as formed according to the equilibrium reaction (1).

$$2R-NH_2+CO_2 \leftrightarrow R-NHCOO^- + R-NH_3^+ \tag{1}$$

The process described here is an enhanced CO_2 absorption based on the Le Chatelier principle, which states that the equilibrium can be shifted by removing one of the constituents in the reaction. In reaction (1), the equilibrium can be shifted towards the formation of carbamate (R-$NHCOO^-$) by removing the protonated amine (R-NH_3^+) from the reaction medium, thus resulting in a higher absorption capacity. For this purpose an amino acid species has been selected. Amino acids react with CO_2 in principle in the same way as alkanol amines, as given by reaction (2).[5] However, amino acids are, in general, more stable than alkanol amines.

$$2\,^-OOC-C(R)-NH_2+CO_2 \leftrightarrow\,^- OOC-C(R)-NHCOO^- +\,^- OOC-C(R)-NH_3^+ \tag{2}$$

Unlike amines, amino acids form electrolytes in aqueous solutions that comprise a zwitterion. Due to its double charge, zwitterions have very limited solubility in water and can be easily separated from the medium by precipitation. In Figure 1, a hypothetical plot is given to depict the difference between a precipitation-based process and a conventional absorption process. This figure shows that for a given partial pressure of CO_2, a process with precipitation has higher loading than a process without precipitation. Therefore, the combined process (simultaneous absorption and precipitating process) leads to additional capacity of the solvent.

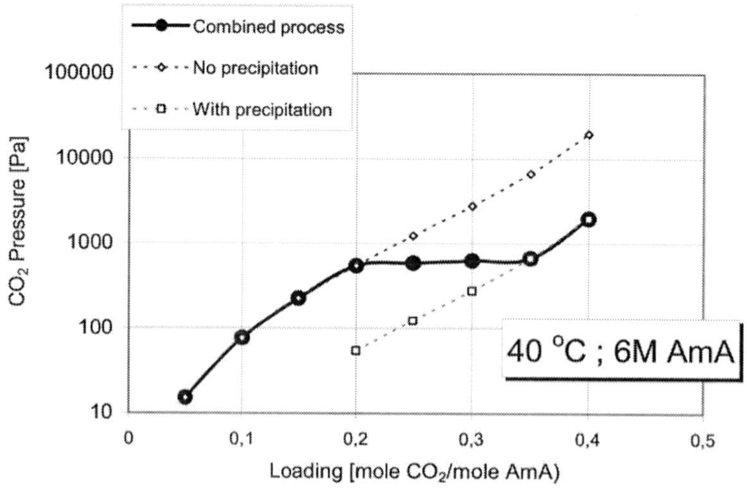

Figure 1 Schematic picture to depict the difference between a precipitating and a non-precipitating system in terms of CO_2 pressure as function of loading.

This works presents the conceptual design of a CO_2 capture process based on the above mentioned principle. The followed strategy combines the use of literature data, estimation of main properties, and a limited number of experiments. The first step in the design, described in Section 1, is the development of process concepts. This consists of selecting and ordering the necessary unit operations that will achieve CO_2 removal from flue gas and recycling of the solvent. Section 2 gives a description of the process and a compilation of technology options for the different unit operations. A structure is provided along with a ranking of options based on selected performance parameters. Section 3 gives a framework of key properties, thermodynamics, and kinetics that allow for the estimation of size and key performance parameters of the process. Based on shortcut methods, a conceptual design of the process is given in Section 4.

1 Process concepts development

The principle described in the previous section requires that a new capture process is designed. Figure 2 shows a block diagram that includes the main steps and functions of the process. The flue gas first needs to be contacted with the solvent (absorption block). The CO_2 then reacts with the solvent, as in reaction (2), resulting in a slurry that contains the carbamate and the precipitated amino acid. Crystals are dissolved (crystal dissolution block) and then the solvent is regenerated (Regeneration block). Finally, the resulting CO_2 product is compressed.

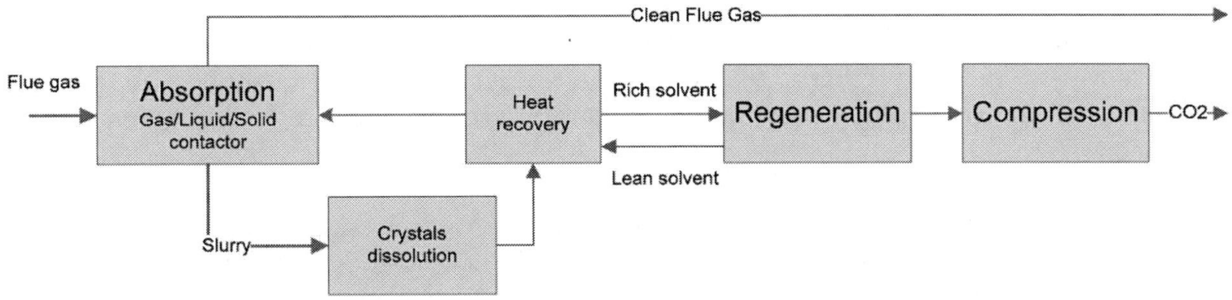

Figure 2 Block diagram showing the structure of the process

The process design allows for various degrees of freedom. The absorption process requires that the three phases (liquid/gas/solid) are contacted. Generally speaking, contactors can employ one of the following mechanisms: dispersion of a gas phase in a liquid or mixture of a liquid and a solid (bubbles), dispersion of liquid droplets in a gas phase (droplets), or contacting a gas phase and a liquid or slurry by means of a fixed contacting surface (surface). Based on these mechanisms, different contactors can be sought out and classified, as in Table 1. Moreover, the regeneration also offers several degrees of freedom (i.e. regeneration at low temperature and low pressure or regeneration at higher temperature and higher pressure).

For the selection of suitable contactors, the following criteria were considered:

1. Fouling behaviour: appropriate for slurry
2. Pressure drop: This parameter influences both process investment (CAPEX) and operational expenses (OPEX) but it has a more important effect in OPEX due to the power consumption to overcome the contactor's pressure drop.
3. Mass transfer characteristics: Mass transfer influences equipment size. This has a direct impact in process investment (CAPEX). In the present study, mass transfer of the different options is represented by the typical operational values of the overall mass transfer coefficient (K_{ol} [m/s])

4. Efficiency: Influences process investment and the purity of the clean flue gas leaving the contactor. This parameter is represented / related to the specific contact area (a [m^2/m^3]), Height of a theoretical plate (HETP) and number of transfer units (NTU).
5. Reliability: based on proven technology, applications in industry, size of operation.

Table 1 Generation of options and ranking for contacting systems. Rating key: **3- Best selection**, 2- Usually applicable, 1- Evaluate carefully, 0- Do not use

Contactor type	Contacting Mechanism[1]	a [m^2/m^3]	$K_{ol} \times 10^4$ [m/s]	Fouling	Pressure drop	Efficiency	Reliability	References
Random Packed column	Surface	100 - 200	0.5 - 2	1	2	2	2	[2],[6], [7], [8]
Structured Packed	Surface	250 - 500	0.5 - 2	1	2	**3**	2	[2], [6], [9]
Tray Column (dual flow)	Bubbles	150	1 - 4	2	1	2	2	[2],[6]
Spray column	Droplets	60 - 100	1 -3	**3**	**3**	0	2	[11],[12],[13],[14]
Multi Stage Crystallizer	Bubbles	20	1 - 4	2	1	2	2	[10]
Bubble column	Bubbles	20 - 400	1 - 4	**3**	1	2	2	[15],[16]
HiGee	Droplets	NA	NA	1	2	2	0	[19], [20]
Hollow fibre module	Surface	3000	0.1 – 0.5	0	1	2	1	[17], [18]

Table 1 shows quantitative estimates for the specific contact area and overall mass transfer coefficient. For the other criteria, the related parameters cannot be quantified at this stage. Instead, a qualitative grading is given based on literature review and the grading system given by Kohl and Nielsen, 1997.[6]

In the absorption case, fouling behaviour, efficiency, and pressure drop are the most important criteria. For the regeneration, efficiency is considered the most important criterion. It follows from this table that in the case of absorption there is no contactor that complies with both fouling and efficiency criteria. For this reason the absorption was divided into two steps and makes use of a spray tower (that handles the slurry when the solvent precipitates) and a packed column (that adds the necessary extra efficiency for 90% CO_2 removal).

In the case of regeneration the packed column is selected due to having a high efficiency, high volumetric ratio, and low pressure drop. In this case, fouling behaviour is relegated to lower importance.

2 Design framework

The recovery of CO_2 from flue gas based on solvent absorption is governed by the phase behaviour of CO_2 and the solvent at the conditions of the absorption unit. For most cases, temperature and pressure are used as operational variables in the regeneration unit in order to release the CO_2 and re-use the solvent. In the present case, the solvent also undergoes a phase change. In order to design the process, the phase behaviour of CO_2 needs to be known at the conditions of the absorption and regeneration units. Moreover, the key parameters influencing the phase change of the solvent (solubility, heat of crystallization, and slurry density) also need to be known. This information has been taken from literature, estimated, or measured.

[1] Surface stands for fixed contact area systems, Bubbles stands for dispersion of gas into a liquid or slurry and Droplets stands for dispersion of liquid into a gas

For the present case, the potassium salt of taurine was selected as the model solvent. The key information necessary for completing the design is listed below:

1. Pure components properties and mixture properties
2. Vapour-Liquid equilibrium of CO_2 with the solvent at absorption and regeneration temperatures
3. Vapour pressure of the solvent
4. Slurry density
5. Mass transfer coefficients realized for the selected contactor type

Table 2 Important pure component properties for Conceptual Process Design

Property	MW [g/mol]	T_m [°C]	T_b [°C]	ΔH_{vap} [kJ/mol]	ΔH_{dis} [kJ/mol]
Taurine	125	305.11	na	na	24.2
KOH	56	420	1327	na	na
Water	18	0	100	40.65	na

Table 2 shows the important pure components information. The Vapour-Liquid-Solid Equilibrium (VLSE) was measured for a 6M solution of potassium taurate. Figure 3 shows a plot of the VLSE of CO_2 at two different temperatures and a concentration of 6M. The same figure also shows the vapour pressure of the solvent mixture at different concentrations. Based on these experiments the solubility of CO_2 can be estimated. Also in these experiments, the precipitation behaviour can be observed and an estimate of the heat of absorption can be given (67kJ/mol CO_2). Sizing the contactors also requires information of equipment mass transfer and reaction kinetics. For shortcut calculations, the overall mass transfer coefficient was determined experimentally for a column packed with Mellapak 350Y and for a spray column under different gas to liquid ratios. Figure 4 shows the overall mass transfer coefficients for these contactor types.

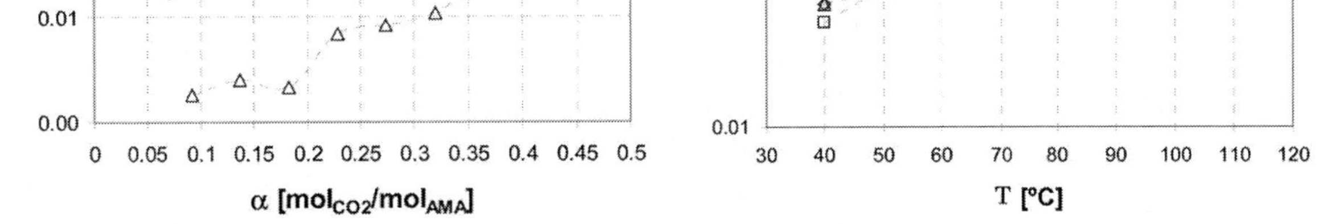

Figure 3 Left – Equilibrium behaviour of CO_2 in a 6M solution at 50ºC and 120ºC
Right – Water vapour pressure of solutions of potassium taurate of 2, 3, and 6M

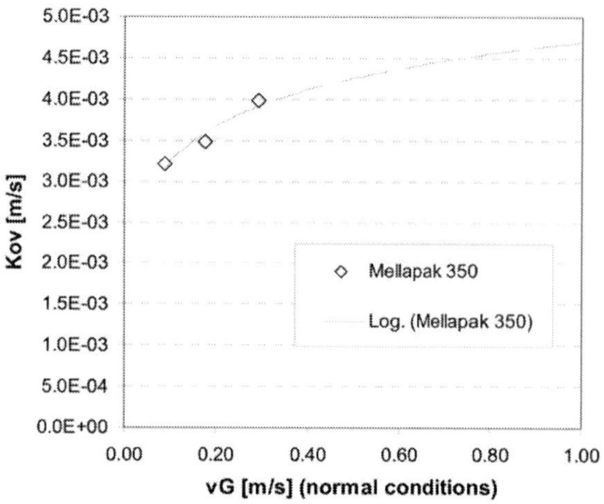

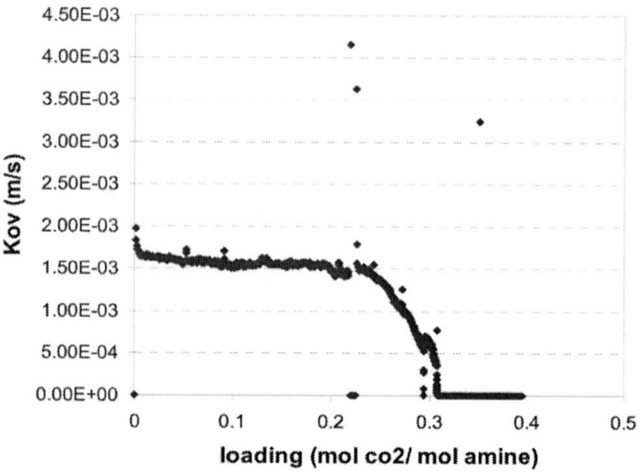

Figure 4 Left - Overall mass transfer coefficient for a column packed with Mellapak 350Y
Right - Overall mass transfer coefficient for a typical spray column experiment

3 Conceptual Process Design

This section presents a conceptual design of the process based on the information given in section 3. Figure 5 shows the process flow diagram of a process that treats refinery flue gas containing 8%vol CO_2 (170 °C, 1.06 bara) and achieve 90% capture. Initially, the flue gas is cooled in a quencher to the absorption temperature (50 °C). After cooling, the flue gas is contacted with pre-loaded solvent in a spray-tower, which consists of a column with no internals. The solvent is sprayed as fine droplets with a high surface area for contacting the gas with the solvent. As a result of this contact, the CO_2 undergoes a chemical reaction with the solvent that leads to the formation of carbamate and carbonate ions. As a consequence of this reaction, the pH of the solvent solution decreases and, with that, so does the solubility of the amino acid. When the solubility limit is reached, the amino-acid precipitates as an amino-acid zwitterion. The resulting slurry is collected at the bottom of the tower with a loading of $0.41 mol_{CO2}/mol_{Taurate}$. The partial pressure of CO_2 in the flue gas is decreased until the point at which massive precipitation of the amino acid occurs. This point has been measured experimentally at different temperatures. Results show that at 50 °C and 6 M potassium taurate, this pressure is 4 kPa CO_2 (0.04 bar CO_2). The remaining CO_2 in the gas is captured in the absorption column, where the depleted flue gas is contacted with lean solvent. The absorption column is a conventional packed column filled in with Mellapak 350Y. There the CO_2 partial pressure is reduced to 0.8 kPa.

The assessment of the technical performance has been carried out by a process model implemented in a spreadsheet application. The aim was to give a rough estimation of the size of the equipment along with mass and energy balances. The most relevant design variables in the process are the temperature and pressure in the stripper (120°C and 1.4 bara) and the gas to liquid ratio in the absorber and spray-tower (542 STPvol/vol). The temperature and pressure in the stripper determine the degree of solvent regeneration and, therefore, the lean loading in the feed solution to the absorber (0.13 molCO_2/molTaurate).

These variables have a direct effect on the percentage of CO_2 removed and the specific energy consumption of the process. Several process calculations were performed until the optimal specific energy consumption was found. The relevant performance parameters corresponding to this point are listed in Table 4.

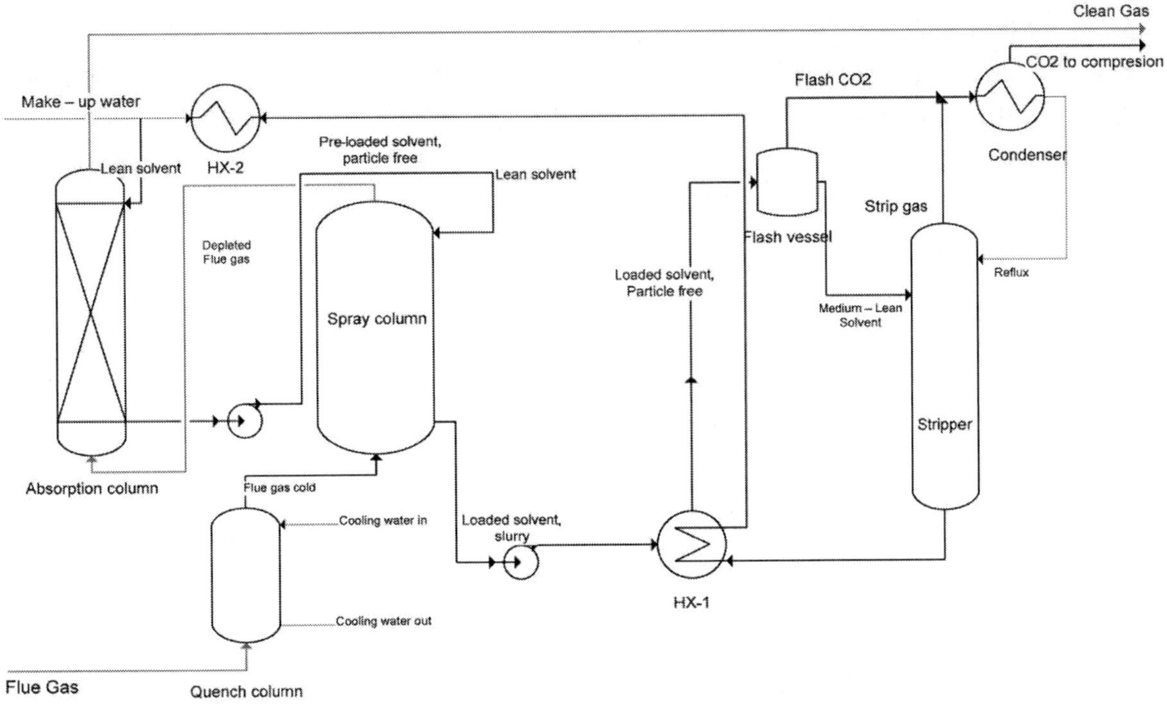

Figure 5 Process flow diagram for capturing 90% of the CO_2 from a refinery flue gas stream containing 8% CO_2

Table 4 Main performance parameters for the process

CO_2 Captured	98295	kg/h
Specific energy consumption	3.2	GJ/ton_{CO2}
Net loading	75	g/L
Contributions to the energy requirement		
Steam production	76135	kW
Crystals dissolution	9955	kW
Total energy	86090	kW
Cooling water requirements	133.11	m^3/ton_{CO2}
Make - up water	0.15	m^3/ton_{CO2}
Electricity	140.81	kWh/ton_{CO2}

The process described has a specific heat consumption of 3.2 GJ/ton_{CO2}. Part of this energy is necessary to dissolve the amino acid crystals and the other part is necessary to regenerate the solvent. However, the initial flue gas could be used to supply the heat duty for dissolving the crystals (10MW). This will result in a reduction of the specific duty to 2.8 GJ/ton_{CO2}. Considering the case of refinery flue gas, which contains less CO_2 than the flue gas from pulverized coal power plants, the given energy consumption is already lower than existing conventional processes.

It is evident that based on the above mentioned results, the DECAB principle can lower the energy consumption for CO_2 removal. This study has shown an example based on potassium taurate. However, by careful selection of appropriate amino acid solvent systems, significant energy reductions can be further achieved.

4 Conclusions

This work shows a process that uses precipitating solvents for CO_2 capture. A new process concept is given that is capable of handling slurries. The key advantages of this process are:

- It is environmentally friendly. Since the process is based on the salt of an amino acid, there are fewer emissions to the air and less degradation products.
- It has low energy consumption due to a lower vapor pressure of the solvent and higher net loading than conventional processes.
- It has the potential to significantly reduce energy consumption by exploring different process options for regeneration and selection of an appropriate amino acid solvent (i.e. with less crystallization energy).

The conceptual process developed allows for an economic evaluation and assessment of process options that will further lead to benefits over conventional processes.

5 References

[1] Danckwerts, P.V., The reaction of CO2 with ethanolamines, Chem. Eng. Sci. (34) 1979. 443-446

[2] Blauwhoff, P., Versteeg, G., Van Swaaij, W., A study on the reaction between CO2 and Alkanolamines in aqueous solutions, Chem. Eng. Sci., 39, 1984. 207-225.

[3] Versteeg, G.F., Duijk, van L.A.J. and Swaaij, van W.P.M., On the kinetics between CO2 and Alkanolamines both in aqueous and non-aqueous solutions. An overview, Chem. Eng. Comm. (144) 1996. 113-158

[4] Barth, D.,Tondre, C., Delppuech, J.-J., Kinetics and mechanisms of the reactions of carbon dioxide with alkanolamines: a discussion concerning the cases of MDEA and DEA, Chem. Eng. Sci. (39) 1984. 1753-1757

[5] Kumar, P.S., Hogendoorn, J.A., Versteeg, G.F. and Feron, P.H.M.., Kinetics of the reaction of CO2 with aqueous potassium salt of taurine and glycine, AIChE Journal. (49) 2003. 203-213.

[6] Kohl A., Nielsen R., Gas purification, Gulf chemical Engineering. 1997 Fifth edition

[7] Metal Random Packing. Sulzer Chemtech specifications

[8] Structured Packings. Sulzer Chemtech specifications

[9] Groserichter, D. and Stichlmair, J., Crystallization fouling in packed Columns, Trans IChemE (81) Part A 2003. 68 - 73

[10] Thieme, C., Sodium carbonates, Ullmann's Encyclopedia 2005

[11] Kuntz, J. and Aroonwilas A., Performance of spray column for CO2 capture application, Ind. Eng. Chem. Res (47) 2008. 145-153.

[12] Warych J. and Szymanowski M., Model of the wet limestone flue gas desulfurization process for cost optimization, Ind. Eng. Chem. Res (40) 12 2001. 2597-2605.

[13] Wijlhuizen A.E., Kerkhof P.J.A.M. and Bruin S., Theoretical study of the inactivation of phosphatase during spray drying of skim-milk, Chem. Eng. Sci (34) 1979 651-660

[14] Wirsching F., Calcium Sulfate, Ullmann's Encyclopedia 2005

[15] Thorat B., Kataria, K., Kulkarni, A.V. and Joshi J.B., Pressure Drop Studies in Bubble columns, Ind. Eng. Chem. Res (40) 2001. 3675-3688.

[16] van't Riet K., Basic Bioreactor Design, Marcel Dekker New York 1991

[17] Zhang, X., Zhang, P., Wei, K., Wang, Y. and Ma, R., The study of continuous membrane crystallization on lysozyme, Desal. (219)2008 101-117

[18] Drioli E., Membrane contactors: fundamentals, Applications and potentialities, Membrane science and technology series. Elsevier (2006)

[19] Lin, C.-C., Liu, W.-T. and Tan, C.-S., Removal of Carbon dioxide by Absorption in a rotating packed bed, Ind. Eng. Chem. Res (42) 11 2003. 2381-2386

[20] Tan C.S. and Chen J.E., Absorption of carbon dioxide with piperazine and its mixtures in a rotating packed bed, Sep. Pur. Technol. (49) 2 2006. 174-1

(92c) A Graphical Method of Improving Batch Reactors

David Ming, Centre of Material and Process Synthesis (COMPS), University of the Witwatersrand, Johannesburg, South Africa, **Diane Hildebrandt**, Center of Material and Process Synthesis, University of the Witwatersrand, Johannesburg, South Africa, Johannesburg, South Africa and David Glasser, Centre of Material and Process Synthesis, University of the Witwatersrand, Johannesburg, South Africa

A graphical technique has been used to determine the production efficiency of batch reactors. We find that by constructing line segments in concentration—time space, between the initial and final states in a standard batch reactor we may determine the production rate directly from the gradient of the line. The gradient of this line is found to be inversely proportional to the production rate corresponding to the batch and hence the batch performance is maximised when the abovementioned gradient approaches its smallest value possible.

We find that interpreting production rate in this manner has identified a new opportunity for improving performance. The opportunity results from the discovery that mixing allows one to overcome areas of low production. As a consequence, the modification is a method of potentially improving production rate in batch reactors by mixing final (desired) product with feed material. The resulting modification shows an improvement over the regular method of operation when no mixing is used when the kinetics is such that the shape of the associated batch profile is concave in concentration—time space. The method is illustrated with industrially significant examples.

See more of this Session: Process Intensification (III)
See more of this Group/Topical: Process Development Division

(92e) Modeling of An Autothermal Catalytic Monolith Reformer to Obtain Hydrogen for Fuel Cells

Safdarnejad Seyed Mostafa and Mohammad Hassan Eikani, Green Research Center and Chemical Engineering Department, Iran University of Science and Technology (IUST), Tehran, Iran

Autothermal reforming (ATR), that is the combination of partial oxidation and steam reforming, is the preferred method for transportation applications because it is a self-sustained process and as a result, increases the energy conversion efficiency and correspondingly reduces environmental impact. The main scope of this work is to simulate the performance of an autothermal catalytic monolith reformer for the production of hydrogen, which is intended to be used in a fuel cell. A three dimensional numerical model was developed based on the computational fluid dynamics (CFD). In the model of monolithic reformer, a single channel was used assuming that each channel behaved the same. The catalysts were deposited on channel walls of the monolith and assumed that the chemical reactions only happened on the catalytic wall surface. The mass transport, chemical reactions and heat transfer can be analyzed simultaneously in the single channel with the aid of CFD simulation. The correctness of the model was verified by comparing the simulation results with the experimental data available in literatures.

See more of this Session: Process Intensification (III)
See more of this Group/Topical: Process Development Division

(92f) Investigation of Traditional and Novel Techniques for Cooling In High Heat Flux Microelectronics Applications

Julaunica Tigner, Mechanical Engineering, Tuskegee University, Tuskegee, AL and Tamara Floyd-Smith, Chemical Engineering Department, Tuskegee University, Tuskegee, AL

The growing demand for electronic devices to be smaller and faster has increased the energy released by these devices in the form of heat. Devices such as laptop computers are not exempted from these demands. The primary traditional technologies currently used to remove heat generated in these devices are fins and fans. Fins dissipate heat through contact with the surface area of a thermally conductive material like aluminum or copper. Fans dissipate heat through air convection. The objective of this study is to perform benchmarking experiments of traditional technologies and to compare the results with more novel technologies like the use of micro channels with water as the cooling fluid. Results are forthcoming.

See more of this Session: Process Intensification (III)
See more of this Group/Topical: Process Development Division

AUTHOR INDEX

AUTHOR INDEX